D1596522

Mobile and Wireless Network Security and Privacy

Edited by
S. Kami Makki
Peter Reiher
Kia Makki
Niki Pissinou
Shamila Makki

Mobile and Wireless Network Security and Privacy

Editors:
S. Kami Makki
University of Toledo
Toledo, OH
USA

Peter Reiher
University of California, Los Angeles
Los Angeles, CA
USA

Kia Makki
Florida International University
Miami, FL
USA

Niki Pissinou
Florida International University
Miami, FL
USA

Shamila Makki
Florida International University
Miami, FL
USA

Library of Congress Control Number: 2007926374

ISBN 978-0-387-71057-0 e-ISBN 978-0-387-71058-7

Printed on acid-free paper.

9 8 7 6 5 4 3 2 1

springer.com

Contents

Preface

Currently the mobile wireless technology is experiencing rapid growth. However the major challenge for deployment of this technology with its special characteristics is securing the existing and future vulnerabilities. Major security and privacy issues for standard wireless networks include the authentication of wireless clients and the encryption and data integrity of wireless LANs. Presently techniques are available to address some of these problems, such as cryptography, virtual private networks. Furthermore the recent advances in encryption, public key exchange, digital signatures and the development of related standards have set a foundation for the flourishing usage of mobile and wireless technologies in many areas such as ecommerce. However, security in a network goes way beyond encryption of data. It must include the security of computer systems and networks, at all levels, top to bottom. It is imperative to design network protocols with security considered at all layers as well as to arm the networks' systems and elements with well designed, comprehensive, and integrated attack defeating policies and devices. A foolproof prevention of attacks is challenging because at best the defensive system and application software may also contain unknown weaknesses and bugs. Thus, early warning systems (i.e. intrusion detection systems) as components of a comprehensive security system are required in order to prime the execution of countermeasures.

As impressive as the theoretical accomplishments of basic network security and privacy research have been, there is still a concern among researchers and practitioners that there is no common and widely acceptable infrastructure in these areas. The need for the explicit organization of such an infrastructure in order to enrich current research and begin the development of practical mobile and wireless networks security and privacy systems that can be widely and easily used, is well understood and accepted by the majority of researchers and practitioners at large. This is self evident from the huge amount of communications which one way or another deal with this subject. For example, the lack of static infrastructure causes several security issues in the mobile ad hoc network (MANET) environment, such as node authentication and secure routing. Even though research in security for MANETs is still in its infancy several security schemes for MANET have already been proposed. Mobile and wireless networking not only complicates routing but security as well. The Ad hoc configurations increase that complexity by an order of magnitude.

This book brings together a number of papers, which represent seminal contributions underlying mobile and wireless security. It is our hope that the diverse algorithms and protocols described in this book will give the readers a good idea of the current state of the art in mobile and wireless security. The authors of each chapter are among the foremost researchers or practitioners in the field.

S. Kami Makki
Peter Reiher
Kia Makki
Niki Pissinou
Shamila Makki

Acknowledgement

This book would not have been possible without the wisdom and cooperation of the contributing authors. Special thanks to the personnel at the University of Toledo and Florida International University and NSF for providing us with a stimulating environment for writing this book.

We would also like to thank Alex Greene, senior Publisher and his staff, specifically Katelyn Stanne at Springer Science & Business for their strong support and encouragements. It was a pleasure working with Alex and Katelyn, who were incredibly patient, very responsible, and enthusiastic about this book. We also would like to express our sincere appreciation to the reviewers of this book, whose suggestions were invaluable.

This book would not have been possible without the indulgence and infinite patience of our families during what often appeared to be an overwhelming task. They graciously accommodated the lost time during evenings, weekends, and vacations. As a small measure of our appreciation, we dedicate this book to them.

1 Research Directions in Security and Privacy for Mobile and Wireless Networks

Peter Reiher[1], S. Kami Makki[2], Niki Pissinou[3], Kia Makki[3], Mike Burmester[4], Tri Le Van[4], and Tirthankar Ghosh[5]

[1] University of California, Los Angeles, CA, USA
[2] University of Toledo, Toledo, OH, USA
[3] Telecom & Info Technology Institute, Florida International University, Miami, Florida, USA
[4] Florida State University, FL, USA
[5] St. Cloud State University, MN, USA

1.1 Introduction

The mobile wireless future is here, and, predictably, the security community isn't ready for it.

Cellphones are ubiquitous, and increasingly have data capabilities in addition to voice, often using multiple different networking technologies. Laptops are in use everywhere, sometimes disconnected, sometimes working off wireless local area networks. Radio Frequency Identification (RFID) is poised to enter our lives, embedded in everyday applications. An increasing number of data appliances of various sorts have become popular, and those of them that are not already augmented with networking capabilities will be soon. Applications are beginning to be built around the very idea of mobility and the availability of wireless networks. And all of these devices and applications are being built for and used by the masses, not just a technologically elite class. As popular as these technologies are today, we have every reason to expect them to be vastly more so tomorrow.

Unfortunately, we are not prepared to secure even the mobile wireless present properly, much less the future. Some technologies and techniques are widely available to help address some problems: cryptography, virtual private networks, and at least the knowledge required to create digital authentication. But these are not nearly sufficient to solve the problems we are likely to face. A few years ago, more or less by accident, the folly of allowing mobile computers to move into and out of an otherwise secure environment became clear, when the Blaster worm used that method to spread into organizations whose firewalls were expected to keep it out. The first worm designed to move from desktop machines to cell phones was recently discovered. The recent cases in Afghanistan of sales in bazaars of stolen flash drives filled with classified data have pointed out that data can be mobile even when full computing and communications capabilities are not. Who knows what other unpleasant surprises are waiting to pop up in this rich, powerful, and poorly understood environment?

The problems are not all unpredictable, either. Providing security for many proposed mobile wireless scenarios is known to be difficult. Mesh networks, and the more mobile ad hoc networks, are known to pose challenges to secure operation that we cannot properly address today. Similarly, the extreme constraints of sensor networks, which usually rely on wireless communications and sometimes feature mobile elements, make many of our standard security solutions infeasible. The scale and openness of proposed ubiquitous computing environments pose tremendous challenges to security. As the available bandwidth and deployment of wireless networks increase, we can predictably expect to see new challenges arise, such as denial of service attacks not easily handled by methods imported from the wired world, stealthy spread of worms by numerous vectors, and clever misuse of the special characteristics of wireless networks for various undesirable purposes.

The same observations are true of the increasingly important issue of privacy. The burgeoning problem of identity theft has made clear that disclosure of private information is not a vague threat only of interest to a handful of activists, but is vital to everyone. The ever growing number cases of disastrous privacy disclosures based on the portability of devices and the openness of wireless networks should make clear that the privacy threats inherent in the wired Internet are going to become much worse in our mobile wireless future. We can so easily lose control of data whose confidentiality we wish to protect when devices holding it are so mobile. And, to a much greater extent than was ever possible before, the presence of ubiquitous wireless networks and portable computers that use them suggests disturbing possibilities for our every move and action being continuously monitored without our consent, our knowledge, or any ability for us to prevent it.

Of particular concern is anonymity and its counterpart, accountability. The loss of privacy and the wholesale surveillance enabled by cell phones, Bluetooth and Wi-Fii capable laptops and devices, as well as RFID tags, affects all of us and may have disastrous consequences. Surveillance, triggered by conflicting interests of companies, corporations and organizations, tracks the electronic footprint of mobile users over network systems, and affects all of us. We urgently need to find simple solutions that give back the user control of their anonymity, while guaranteeing accountability.

One important aspect of securing the wireless mobile future that must not be overlooked is that it will be a future of the everyman. The users will not be elite, will not be security (or even networking) specialists, will not be willing to learn many new skills to make use of their devices, and will not have regular access to trained security and system administrators. The security for this future world cannot depend on complex manual configurations, deep understanding of security threats by typical users, or reactions to ongoing problems by the humans working with the system. One of the most consistent lessons of computer security technologies is that only the technologies that are invisible to the average user are widely used. We cannot require any significant setup by the average user, we cannot require ongoing human monitoring of the behavior of the typical device in this environment, and we cannot expect user-initiated reactions to either potential or actual threats. Anything that is not almost completely automatic will not be used. If we look ahead to the predicted ubiquitous computing and sensor network future, this observation becomes even more critical. There will not be a security professional monitoring and adjusting the behavior of smart wallpaper in the typical home or vast undersea sensor networks monitoring the ocean's floor for seismic activity. We must move to a future where these devices and networks are secure on their own, without ongoing human supervision.

So the computing world is already mobile and wireless, and is becoming even more so rapidly and unalterable. And we cannot even secure the relatively simple environment we see today. These dangers motivated the National Science Foundation to fund this study of the requirements for research in the field of mobility and wireless networks. The study is based on the deliberations of a group of leading researchers in this field at an NSF-sponsored workshop on security and privacy for mobile and wireless networks (WSPWN), held in March 2006 in Miami, Florida. This workshop presented position papers on the threats and possible mechanisms to handle these problems, which lead to deep discussions by the participants on what was lacking in the current research in these areas, and where the National Science Foundation and other agencies able to fund and direct research should try to focus the research community's efforts. This report distills the results of that workshop.

The report opens by presenting a brief view of the current situation in the fields of privacy and security for wireless and mobile networks, covering both the knowledge we have available already from existing research and the range of threats we have seen and can predict. The report goes on to discuss areas where the workshop participants agreed more research was vital. We also discuss the general character of the kinds of research we feel is more necessary and elements that funding agencies should look for in research proposals in this area.

1.2 The State of the Art

All is not totally bleak in the field of security and privacy for mobile and wireless networks. We can start by inheriting a number of useful tools from other fields, and some good research has already been done in certain vital areas, sometimes leading to techniques and tools that will certainly help us solve many future problems. On the other hand, there are many open problems and unaddressed needs.

To begin with the brighter side of the picture, much of the work already done in cryptography has a great deal to offer wireless networking. Cryptographers have always preferred to work on the assumption that their opponents can both overhear and alter the contents of the messages they send. In wired networks, doing so was often difficult. In wireless networks, it's usually easy. Since the encryption algorithms and cryptographic protocols tend to take such effects into account, they are still perfectly usable in the wireless domain. So we already know how to maintain privacy of data sent out over wireless networks, how to detect improper alterations of such data while in flight, and how to determine the authenticity of messages we receive over wireless networks. This is not to say that all cryptography-related problems related to wireless networking have been solved, but we do at least have solid knowledge that can be used to build working tools right now, and that can serve as a basis for solving other security problems.

Unfortunately, as has been proven time and again in wired networks, cryptography alone cannot solve all security problems. So the mere presence of good encryption algorithms and cryptographic protocols does not always take care of our difficulties. For example, many devices that use wireless networks are powered by batteries. Often, as in the case of sensor networks, these batteries are quite limited. For that matter, the devices themselves might have other strong limitations, such as limited processing capacity, memory, and secondary storage. Much of the best cryptography requires significant amounts of computing. Years of research on cryptography for low power devices has not yet succeeded in finding algorithms that we can regard as being as secure as those that are usable in less constrained circumstances, nor techniques that can convert existing algorithms into low powered variants with little or no loss of security, although some recent results on "light" cryptography are promising

Cryptography also has something to offer for mobile devices. The rash of recent cases of lost or stolen laptops and flash drives holding sensitive information should have taught security-aware users that the sensitive data they store on these devices should ordinarily be kept in encrypted form. Even when they do keep such data in this form, however, they still must decrypt the data before they can use it, which opens a number of possibilities for mobile devices in dangerous environments failing to protect their sensitive data based on cryptography alone. Some research has already been performed on ensuring that only the mobile device's authorized user can get to its data. Much more needs to be done. And we should never forget one critical fact about cryptography: it simply reduces the problem of protecting data to that of protecting cryptographic keys. If keys are stored insecurely or users can be fooled into providing them when they shouldn't, the potential security offered by cryptography fades away. And unless secure key recovery measures are taken, the loss of the keys results in the loss of stored data.

Other existing security technologies, such as firewalls, have something to offer. While the traditional deployment of firewalls at the (virtual) point where a network cable enters an organization's property has been shown to be inadequate in wireless and mobile environments, the idea of a perimeter defense between a network and a computing capability still has some value. The most common wireless networks (both cellphones and 802.11 LANs) usually work in an access point mode, where communicating devices always send their data through an access point, even when the receiver is in direct radio range. This access point is a natural location to put a perimeter defense, and a number of vendors offer capabilities of this kind. In the wired mobile computing case, the lessons of the Blaster worm have led to some simple firewall-like technologies being applied whenever a device is first connected to the network, at least until that device has been determined to be free from the most obvious and dangerous threats. Personal firewalls that protect a single computer (typically a portable computer) from threats wherever it is and whatever networking technology it is using are generally available and are often fairly effective. This reduces the problem of securing mobile devices to the more manageable problem of securing access points.

Other existing security technologies are still applicable to the mobile and wireless environments. Prosaically, but importantly, methods used to evaluate the security of wired environments can be extended to evaluate the security of wireless ones, provided those doing the extension understand the special characteristics of wireless networks. Auditing and logging retain their value in the wireless mobile world. Many forms of two-factor authentication already expect a human user to carry a card or a device with him to assist in authenticating him, and that paradigm is likely to work equally well when the user moves from place to place. Tools that are intended to work purely on a single machine, like virus detection software, will generally be useful for mobile single machines as much as fixed ones.

However, even intelligent application of these and other useful technologies does not cover all the security problems of the mobile wireless world. The remainder of our report will concentrate on areas where we see a need for further research.

1.3 Areas for Future Research

1.3.1 Challenges for standard wireless networks

1.3.1.1 802.11 Wireless Networks (Wi-Fi)

Wireless networks have experienced an explosive growth because of their significant advantages of productivity and convenience. A major challenge for deployment of this technology is securing its new vulnerabilities. All too often, such networks have been deployed without any thought of such challenges, often leading to security disasters. Major security issues for standard wireless networks include the authentication of wireless clients and the encryption and data integrity of wireless LAN frames, as analysts believe that the wireless LANs can be easily accessed by outsiders (friendly or not) and need strong protection.

The IEEE 802.11 standards, often called Wi-Fi (wireless fidelity), are the family of wireless specifications for managing packet traffic for multiple users over a wireless network. These standards were developed by a working group of the Institute of Electrical and Electronics Engineers, and have achieved wide popularity in enterprise, home, and public settings. Although a number of security measures were built into the 802.11 standard, such as the Wired Equivalent Privacy protocol (WEP) and Wi-Fi Protected Access (WPA), it is almost universally accepted that wireless networks are considerably less secure than wired ones. Some of the problems leading to such insecurity are inherent in the very idea of wireless networking, some are specific to the style of wireless networking supported by 802.11, and some are caused by particulars of the protocols specified in these standards.

A wireless network uses signals such as light or radio waves to provide connection among the different devices such as computers, phones, etc. Therefore, wireless networks share airwaves with each other, and the radio signals typically travel in all directions. Technologies using directional antennae and relatively tight beams, such as some free-space optical systems, limit the area in which an attacker can access the transmission, but for the more popular technologies, anyone within the range of a wireless network can access or intercept an unsecured system. Therefore, hacking into a wireless system can be simple if the standard security features such as encryption are not in place. These measures, when added, only protect data from the user end point to the wireless access point; from that point on, the data will be unencrypted and passes in the clear. A well-established guideline is to treat the wireless LAN as an untrusted network, like the Internet, and to install a firewall or gateway where wireless and wired networks meet.

Even when in place, these measures are far from perfect, since they provide only the elements of security that encryption can provide. Thus, they do little for handling denial of service, they are of limited value for any attack that relies on traffic analysis, and they do not necessarily protect the network from misbehavior by those who have some degree of legitimate access. These are areas of concern that merit further research.

Wireless technology has already proven extremely useful, and holds even greater promise, but it also poses great technical challenges. Recently, Meru Networks has proposed a software solution for protection of wireless networks at the Radio Frequency (RF) level. They propose microscanning, radio scrambling, and transmission jamming of the radio waves in order to ensure a fine level of security for any enterprise. Approaches that leverage the characteristics of wireless transmissions in general, and the specific characteristics of the bandwidths in popular use, are a fertile ground for further research.

As more companies and individuals make use of wireless applications, protecting privacy and confidentiality will be paramount. Therefore, well-designed solutions for securing, mobilizing and managing wireless LANs should integrate seamlessly into existing enterprise network design and network management principles. At the moment, the technologies for supporting such integration are not highly developed. Research in this area would thus be of great value to many people and organizations.

1.3.1.2 3G Wireless Networks

The popularity of cell phone technology and Wi-Fi networks has led to development of further wireless technologies to allow easy data transmissions to and from various devices, especially cell phones. These technologies are often called third generation, or 3G, wireless networks. Various standards and systems have been built around 3G concepts, which are widely deployed in some countries and are expected to achieve popularity in many others. The most significant features offered by third generation technologies are huge capacity and broadband capabilities to support greater numbers of voice and data transfers at a lower cost. The rapid evolution of 3G technologies has provided the ability to transfer both voice and non-voice data at speeds up to 384 Kbps.

Having learned some lessons from the difficulties early 802.11 systems had with security, and because of the increasing government and standards body requirements to protect privacy, security played an instrumental role in the design of 3G technologies. However, 3G wireless networks not only share all kinds of wireless networks vulnerabilities, but also have their own specific vulnerabilities, such as stealing cellular airtime by tampering with cellular NAMs (numeric assignment numbers).

Further, 3G technologies are likely to operate side by side with other forms of wireless networks. Therefore, organizations, both public and private (such as the Third Generation Partnership Project, or 3GPP), are exploring ensuring safe and reliable interoperability of 3G and wireless LAN technologies. One of the main problems that threaten this interoperation is the lack of thorough and well-defined security solutions that meet the challenges posed by the combination of these technologies. Further research is required in this area.

While the most obvious threats to 3G and other wireless network technologies are active attacks on the radio interface between the terminal equipment and the serving network, attacks on other parts of the system may also be conducted. These include attacks on other wireless interfaces, attacks on wired interfaces, and attacks which cannot be attributed to a single interface or point of attack. Better understanding of the range of such attacks, methods of designing networks less susceptible to them, and countermeasures to protect systems being attacked in these ways are all valuable areas of research that NSF should support.

Generally, the introduction of any new class of wireless network into either common or specialized use also introduces the possibility of attacks on its special characteristics and attacks on the points at which the new class of network connects to or interacts with existing networks, wireless and wired. Any networking research that the NSF supports on new classes of wireless networks should be complemented with security research that addresses these threats. There is no point in repeating the mistakes made in securing 802.11 networks, and great value in learning from the good examples of designing security into 3G technologies.

1.3.2 Challenges for sensor networks

Advances in technologies such as micro-electro-mechanical systems (MEMS), digital electronics, and the combination of these devices with wireless technology have allowed information dissemination and gathering to/from terrains that were difficult or impossible to reach with traditional networking technologies. Today's sensors are tiny micro-electro-mechanical devices comprise of one or more sensing units, a processor and a radio transceiver and an embedded battery. These sensors are organized into a sensor network to gather information about the surrounding environment. Both the sensors and the sensor network are commonly expected to be largely self-managing, since many proposed uses require deployment of large numbers of sensors in remote or inaccessible areas, with at most occasional attention from human beings. The self administering properties of sensor nodes and self organization of sensor networks, combined with random deployment features, allow them to be used for a wide range of applications in different areas such as military, medicine, environmental monitoring, disaster preparedness, and many others.

Because of the limited power of sensor nodes, their specialized purpose, and their need to be almost entirely self-administering, a new class of network protocols and designs has been developed for sensor networks. They do not have the same capabilities, needs, or purposes as a typical networked computer, even a typical computer that uses wireless networking. As a result, security solutions developed for the Internet, wireless LANs, or other more standard purposes are often either unusable or irrelevant for sensor networks.

The use of sensor networks in mission-critical tasks, such as allowing the military to monitor enemy terrain without risking the lives of soldiers, has demanded urgent attention to their security, and has thus been the focus of many researchers. While the lower level characteristics of the network and its capabilities are very different, at a high conceptual level the provision of the security in this environment has the same requirements as any other network environment: confidentiality, data integrity, data freshness, data authentication and non-repudiation, controlled access, availability, accountability, etc. Important research must be done, however, in matching these security requirements to the specific needs and limitations of sensor networks. Examples of special security problems for sensor networks include:

- Cryptography and key management – The sensor nodes usually have very limited computation, memory, and energy resources. Symmetric cryptography algorithms face challenges in key deployment and management, which complicates the design of secure applications. On the other hand, asymmetric cryptography's higher computational and energy costs render it too expensive for many applications. In many cases, the particular needs of sensor node applications suggest that lower levels of protection are acceptable than in other networks. For example, much data gathered by sensor networks is time critical, and its confidentiality need only be protected for some limited period. Matching the style and costs of cryptography to the needs of particular sensor networks is an important problem for research.

- Node integrity – In many cases (including critical military scenarios), sensor networks must be deployed in areas that are readily accessible to opponents. Thus, sensor nodes can be easy to compromise due to their physical accessibility. The compromised nodes may exhibit arbitrary behaviour and may conspire with other compromised nodes. Designing sensor network protocols that are tolerant to some degree of node compromise is one important area of research. Another is designing suitable methods of detecting compromised sensor network nodes and securely reconfiguring the network and application to avoid them.
- Scalability - Sensor networks may have thousands or more nodes, requiring consideration of scaling issues. Some security techniques are not designed to operate at all at the scale sensor networks will exhibit, and others will have increasing costs at high scale that cannot be born by sensor networks. Research is needed on understanding the scaling costs of security algorithms, studying the effects of those costs on sensor networks, and designing high scale security solutions specific to sensor networks.

Due to inherent limitations and requirements of sensor networks, a number of different and new security mechanisms, schemes and protocols need to be created. Different attacks on sensor networks can occur in different network layers (physical, data link, network, and transport). For example, at the physical layer an attack can take the form of jamming the radio frequency or tampering with the nodes of the network. At the data link layer, attackers can exploit collisions, resource exhaustion, and unfairness. At the network layer, attacks can include spoofing, data alteration, replays of routing information, selective forwarding, sinkhole attacks, white hole attacks, sybil attacks, wormholes, HELLO flood attacks, insulation and corruption attacks, or acknowledgement spoofing. At the transport layer, the attacks include flooding and desynchronization.

Popular security approaches in sensor networks can be classified as cryptography and key management, routing security, location security, data fusion security, and security maintenance.

- Cryptographic concerns that are particularly important for sensor nets include the processing and power costs of performing cryptography, complexity of the algorithms (since sensor network nodes often have limited memory to store programs), and key distribution. In addition to the normal problems with key distribution for any network, sensor network nodes try to minimize network use, since sending and receiving messages drains battery power. Key distribution is thus competing with the core purpose of the sensor network for a scarce resource, and must therefore be designed carefully.
- In many sensor networks, routing protocols are quite simple and offer few or no security features. There are two types of threats to the routing protocols of sensor networks: external and internal attacks. To prevent external attacks, cryptographic schemes such as encryption and digital signatures can be use. However, internal attacks are harder to prevent, since detecting malicious routing information provided by the compromised nodes is a difficult task. Techniques developed for this purpose for other types of networks, such as ad hoc networks, often rely on sharing information among many nodes or performing complex analysis on information gathered over the course of time to detect potential cheating. Sensor networks' special resource constraints might make such techniques unusable. On the other hand, sensor networks typically use very different styles of routing strategies than other types of networks, and it might prove possible to leverage those differences to achieve some security goals. More research is required here.
- Location security is important when the proper behavior of a sensor network depends on knowledge of the physical location of its nodes. While sensor network nodes are not usually expected to move (for a wide range of sensor network applications, at least), they are often small enough and accessible enough for malicious entities to move them as part of an attack. Being able to tell where a sensor network node is located can often have important benefits, and, conversely, attackers may gain advantage from effectively lying about locations.
- Data fusion is a normal operation to save energy in sensor networks. Rather than sending each node's contribution to the gathered data to the data sink, data is combined and forwarded. However, if some sensor network nodes are compromised, they can falsify not only their own contribution, but any fused data that they are supposed to forward. Standard authentication techniques do not help. Alternatives include collective endorsements to filter faults, voting mechanisms, or statistical methods. Another approach is to use data aggregation methods that can work on ciphertext in intermediate nodes.

- The detection of compromised nodes and security maintenance also are important. In some methods, the base station gathers information from sensors and processes it to find compromised nodes. In other methods, neighboring nodes cooperate to determine which nearby nodes are behaving badly. Other methods are integrated with the particular application to detect security faults. In some cooperative approaches, statistical methods or voting methods have been used to find the compromised nodes.

Sensor networks are usually considered to consist of active, battery-operated nodes. However, another class of wireless networks that perform sensing uses passive or reactive power-free nodes. One example is a network designed to interact with RFID tags. Although readers are needed to power-up the sensors, the deployment life-cycle of such systems has no apparent limits. This seems to be a very promising area for some applications, and can be used very effectively to manage power resources. However, some of these passive technologies have some very serious security concerns, and more research is required to understand how they can be safely integrated into systems with strong security requirements.

Other forms of more exotic sensor networks might include robotic mobile nodes or close interactions with more classic forms of wireless networking. These forms of sensor networks are likely to display new security problems, and, conversely, offer interesting security opportunities based on their unique characteristics.

1.3.3 Challenges for mesh and ad hoc networks

Mesh and ad hoc networks offer the possibility of providing networking without the kind of infrastructure typically required either by wired networking or base-station oriented wireless networking. Instead, a group of wireless-equipped devices are organized into a multihop network to provide service to themselves. Sometimes, the mesh or wireless network connects to more traditional networks at one or more points, sometimes it stands alone as an island of local connectivity in an otherwise disconnected area. The primary difference between mesh and ad hoc networks is usually that a mesh network tends to have less mobile nodes, and thus the network connections established tend to persist for a long period, while an ad hoc network typically assumes frequent mobility of some or all of its nodes, meaning that the set of nodes reachable from a particular wireless device changes frequently.

For the purpose of this report, we care about the privacy and security challenges of these networks only. However, it is worth noting that it is unclear whether the basic networking challenges of these types of networks have been sufficiently solved to make them generally attractive, regardless of security issues. To the extent that we are unsure of the fundamental methods to be used to provide networking in this environment, such as which algorithms will be used to find routes between nodes, it might be hard to determine how to secure the networks. But some aspects of security are likely to be common for all networks of these styles, and it behooves us to address security and privacy challenges of these kinds even before the basic networking methods have been worked out.

There are clear security challenges for these networks. Beyond those inherited from using wireless at all, the core idea of ad hoc networking requires cooperation among all participating nodes. Unlike even standard wireless networking, there is no permanent infrastructure, and no particular reason to trust the nodes that are providing such basic network services as forwarding packets. All nodes send their own messages, receive messages sent to them, and forward messages for other pairs of communicating nodes. Routing protocol security based on trust in the routers (which is really the paradigm used to secure Internet routing today) does not work too well in ad hoc networking. Further, the assumption of high mobility typically also implies that all participating nodes are running off batteries. Attacks on the energy usage of nodes are thus more plausible than for other types of networks. Also, since radios have limited effective ranges and generally ad hoc networks are intended to span larger areas than their radios can cover, issues of the physical locations of nodes might well be important, leading to new classes of attacks based on obtaining privileges or advantages by lying about one's location.

In addition to needing to provide routing from normal, possibly untrusted peer nodes, mesh and ad hoc networks will have to rely on such nodes for all other services. For example, if a DNS-like service is required by the network, some peer node or set of such nodes will have to provide it, since there is no one else to do so. Therefore, the lack of trust that a node can have in its service providers extends up the networking stack, all the way to the application layer. More research is probably warranted in providing security for ad hoc and mesh network services beyond routing. At least DNS services, and possibly quality-of-service mechanisms and proper behavior of administrative protocols like ICMP should be examined in the context of these specialized networks. Some understanding of how to design distributed applications for such an environment also warrants research attention.

One outcome from existing research on ad hoc networks seems to be that achieving the same level of basic network service in such environments as in even a standard access-point based wireless environment is very challenging. This suggests that ad hoc networks are most likely to be used when standard networking methods are out of the question. The most commonly suggested scenarios for ad hoc networks are military (when a unit needs to operate in an area with no existing networking infrastructure, or is unable to use the existing infrastructure), disaster relief (when previously existing infrastructure has been destroyed by the disaster), and critical infrastructure protection (for overlay or backup (sub)networks). It might be beneficial, given the likely difficulties of securing such complex networks and the inability of researchers to identify many other promising uses for ad hoc networks, to concentrate on the particular security requirements of these scenarios.

1.3.4 Challenges related to mobility

While mobile computing is often considered in conjunction with wireless networking, it is not really the same thing. Many mobile computing scenarios do not involve any wireless communications whatsoever. The worker who unplugs his laptop computer from his office Ethernet, drives home, and plugs it into his home DSL router has performed mobile computing, for example. Thus, some security and privacy issues related to mobile computing are orthogonal to many of wireless issues.

One key issue for mobile computing that has been underaddressed is the unglamorous issue of theft of these devices. Mobile computing devices are, almost by definition, relatively small and light. Also, they are taken to many places. As a result, they are often stolen. In addition to the incidents of stolen flash drives in Afghanistan, we have seen increasing trends towards "snatch and grab" crimes against laptop computers in cybercafés. There have been many serious privacy breaches related to precisely such incidents. When a laptop computer carrying private information is stolen from an airport, a coffee shop, or a bus, the data it carried becomes at risk. In many cases, the owner and his organization might only have a vague idea of what information is actually on that lost laptop, and thus the magnitude of the theft. Conventional wisdom suggests that merely applying standard cryptography to the file systems of mobile computers will solve the problem, but we have heard many times before that mere use of encryption will solve a problem. Often, the practical use of cryptography is more complex than it seems at first sight. Some likely complexities that should be addressed relate to key management for this environment (if the keys encrypting the data are stored on the machine in readily available form, the cryptography is of little value), usability (encrypted data is of limited use), and purging all traces of the unencrypted form of the data (caching is widely used at many levels in modern computer systems). Similarly, the simple claim that security cables will solve the problem requires closer examination, since, in many portable devices, the disk containing the vital data can be easily removed from the device.

A separate technological development, the increasing size of disk drives, has made it common for data once placed on a machine to remain there forever, since the disk is large enough to handle the storage needs of the user for the lifetime of the machine. (Perhaps not when talking about huge media files, but few people bother to clear out documents, spreadsheets, or electronic mail messages to make space.) Thus, either a human user remembers to clear private data off a mobile device when he is done with it, or it remains there forever. Should the device be stolen or discarded, a vast amount of such data might go with it. Can technology offer any assistance to solve this problem? Should some automated system delete, move, or encrypt old, unaccessed data on a laptop computer? If so, how, which data, and when? If it is deleted, how can we be certain we haven't lost vital data? If it is moved, where to? If it is encrypted, with what key, and how does the user recover it if needed? How is stored data protected on devices made obsolete through technology advances?

Another important question for mobility is that mobile computers can enter environments that are not under the control of their owner, nor under the control of people that the owner trusts. A desktop machine can be protected by a company's IT department. A laptop is only so protected until the owner walks out the door of the company's office building. From that point onward, it becomes a visitor in strange and unexplored realms potentially filled with unknown perils. Why should the user trust the coffee shop or the Internet cafe that offers him access? How can he be sure that the small hotel that throws in free network connectivity with its room rate is sufficiently well secured? What (other than not connecting to a network at all) can he do to achieve the degree of security his needs demand?

One wonderful possibility offered by mobile computing is that a group of users who happen to congregate together in a physical place can use their devices (probably over a wireless network, but not necessarily) to interact. They can:

- share their data
- pool their computing, storage, and communications resources
- set up temporary applications related to their joint presence in a particular place for a particular period of time
- learn about each other
- foster social interactions in many ways

Even if they have little or no connectivity to the greater Internet, they can still share rich communications experience via simple ad hoc networking or through setting up a local wireless “hub” in their environment.

This possibility sounds very exciting. However, to a security professional, it also sounds like a recipe for disaster. My computer is going to connect to the computers of a bunch of near-strangers and allow them to do things? What if they’re villains, or, almost equally bad, incompetents? How can I limit the amount of damage they are capable of doing to my precious computing environment, while still allowing useful social interactions?

This problem is magnified when we consider the postulated ubiquitous computing environment of the future. In this vision, while potentially many of the computing devices in the environment could communicate to the Internet, most of their functions are intended for physically local consumption, and they are often designed specifically to meet the needs of mobile users passing through the physical space they serve. These ubiquitous devices are thus expecting to interact with large numbers of users they might never have seen before, and might never see again, for perhaps relatively brief periods of time. The environment must also protect itself against malicious users who wish to disable it or use it for inappropriate purposes. By its nature, these protections cannot be strong firewalls that keep the attackers out, since generally an attacker can move into their physical space. Once he does so, unless he can be identified as an attacker, he seems to be just another user who should get service from the ubiquitous environment. What can the environment do to protect itself from the bad users while still offering rich services to the good ones? Turned on its head, the question becomes what can a mobile user moving through various ubiquitous environments do to make safe use of the services they offer, while ensuring that malicious or compromised ubiquitous environments do not harm him? One particular aspect of this latter question relates to location privacy. In a ubiquitous future, where people usually carry computing and communications devices wherever they go, and those devices typically interact with ubiquitous computing installations at many places, how can a user hope to prevent information about his movements from becoming public knowledge? Must he turn off his useful communications devices if he wishes to retain privacy, or can he make some adjustments that allow him to use them while still hiding his identity from the environment, or otherwise obscuring his movements? If we combine this issue with the earlier one of ensuring responsible behavior by users in ubiquitous environments, we see a serious concern, since one way of preventing misbehavior is detecting it and punishing the malefactor. Yet if users of ubiquitous environments can hide their identities, how can we even figure out who was responsible for bad behavior?

A related issue deals with user control of private data. Nowadays, if a user trusts the Internet, but does not trust wireless networks, there is no way for him to determine if messages he is sending containing private data will or will not cross networks he doesn't trust. Is there a practical way for users to control such data flow, limiting it to only sufficiently trustworthy portions of the network? Similarly, is there any way for a user to force his private data to be kept off portable devices, or to be stored only in encrypted form in such devices?

Another interesting security research question is how to formulate a trust model in this ubiquitous and dynamic environment. Mobility creates huge problems in formation of such a trust model, which is even more difficult when near-strangers are required to communicate without having any past communication experience. The trust formulation must take into account the possible malicious behaviors of the participating hosts without merely concentrating on parameter collection from previous experience.

In many cases, the degree and patterns of mobility that actually occur might have a strong effect on the security of mobile devices. A trivial example is that a laptop that is only moved from one desk to another in a secured facility is at less risk than a laptop that is carried on travels all around the globe. A more complex example relates to location privacy. While clearly a location privacy solution that works equally well for any possible movement pattern is best, such solutions might prove impossible to design. In that case, a solution that works well for the movement patterns observed in the real environment where the technology is to be deployed is the next best thing. Similarly, when analyzing the kinds of risks devices face as they move from place to place in a ubiquitous environment, the style and pattern of that movement might have a significant effect. Some, but relatively little, data on real movement of users in wireless environments has started to become available, but more is needed, both for general mobility research and for mobile security research. In addition to raw data, we need realistic, but usable, models of mobility to drive simulations and to test new mobile security technologies. We are likely to need models for many kinds of uses, since a model that properly describes the movement of cars on a freeway is unlikely to also accurately describe the movements of customers in a shopping mall or workers in a factory.

1.3.5 Security for new/emerging wireless technologies

We can expect that researchers will continue to develop new wireless technologies. Some will be designed for special purposes, such as underwater sensor networks. Others will take advantage of changes in spectrum allocation that open up new bandwidths for more public use. Others, like free space optical networks, are already under development, though it is not yet clear how widely and in what modes these might be deployed and used.

To the extent that these networks are truly different than existing forms of popular wireless networking, we can predict that the security challenges (and opportunities) related to their use will also be different. The National Science Foundation should urge researchers in network security to keep abreast of new developments in wireless networking technology and to consider how to meet their new challenges before such systems are completely designed and start to be deployed. Similarly, as networking proposals based on novel technologies are being considered for funding, the NSF should always insist that the networking researchers deeply consider the privacy and security implications of their work. No one will benefit from repeating the security mistakes in the design of 802.11.

1.4 General Recommendations for Research

Some might object that much research funding has already been poured into the field of wireless networks and mobility, often with few practical results to show for it. There are certainly many papers describing ways to provide security for a variety of protocols for handling routing in ad hoc networks, for example, but we have few practical ad hoc networks, and little or no experience with actual attempts to attack them or successes by these technologies in countering such attacks. But if one accepts the rather obvious fact that we have no secure wireless ad hoc networks, and that other areas in wireless and mobile systems that have received much study are in similar condition, one must then try to identify what elements of the earlier research failed to lead to solutions to these problems. Are they merely early steps that, while they have not yet borne much fruit, need nothing more than persistence? Or are there fundamental problems with the directions that have been taken, requiring a fresh start based on the lessons we've learned from the limited success of existing methods?

There is at least one fundamental problem with much of this research: all too often it is not based on reality. A great deal of research in the mobile and wireless security arena (and, for that matter, all forms of mobile and wireless research) is based purely on analysis and simulation. Many algorithms and systems are never implemented in real environments at all, yet they go on to become well known and highly cited. The methods they used are adopted by others, and an elaborate edifice of research is built on what must inherently be a flimsy foundation. This observation is particularly true because the common experience of those who have worked with real deployments of mobile and wireless networks have discovered that they tend to be unpredictable, changeable, and hard to characterize, all particularly bad characteristics when relying on simulation or analysis. Those techniques work best for well understood phenomena where it is reasonable to create models that are close approximations to observed reality. In the wireless realm, the reality observed is not often like the models used in much research. Similarly, the models of mobility used in such research are too simplistic and have no grounding in actual behavior of users and other mobile entities.

While this observation is unfortunately true, it should not be regarded as a harsh criticism of those who have done this research. Most of this research was done when the mobile and wireless environment really was the future, not the present. There were often few or no suitable networks to test with, and their character and mode of use could only be predicted or speculated on. Early researchers in this field had little choice but to rely heavily on simulation and, to a lesser degree, analysis.

But that time has passed. There is no great barrier today to creating a wireless network, at least one based on commonly used technologies. Almost all laptops come with one, two, or even three different forms of wireless communications built in. Because people have actually become mobile computer users, there is no further need to speculate about either how they will move or how they will behave when they move. They are out there doing it, in large numbers, everywhere we look.

Therefore, a major recommendation of this report is that future research in security and privacy for mobile and wireless environments should be performed on a basis of realism. Simulation should be used as a supporting method of evaluating a system, not as the only method. More attention should be paid by researchers to the realities of what is actually happening every day, rather than relying on outmoded models that were created when it was only possible to guess what might happen. Most research should result in working prototypes. Most research should make use of either live tests or modeling based directly on observed behavior of real users and systems working in the targeted environment. While the NSF cannot abandon deep theoretical research or early investigations in new areas, more emphasis should be placed on solving the privacy and security problems we already know we have and cannot solve in real networks that are in use today.

This recommendation is not solely based on researchers' obligations to perform their research in the most intellectually defensible method possible. It's also based on pure practical necessity. The mobile and wireless environment is not secure now, and will not become much more secure unless research is done into suitable ways to achieve that goal. In an era of limited available funding for such research, priority must be given to work that will improve the security of the systems we see in use today and that we can definitely expect to see deployed in the near future. Some resources must still be directed towards theory and development of revolutionary technologies, but the needs of the present should not be neglected by researchers. To the extent that NSF priorities influence the agenda for many researchers, directing their attention towards important problems in today's wireless mobile environment for which we do not even have promising research directions is important.

This recommendation of realism extends further than merely favoring system development and real world testing. It also extends to the areas that should be funded. The NSF should encourage research that address security problems that are being actively exploited today, projects that help us to better understand the actual uses of mobility in the real world, and the actual behavior of wireless networks in real environments. Tools that help researchers build and test their privacy and security solutions for such realistic environments would be valuable.

This argument is not intended to shut down theoretical research or bold research that seeks to move far beyond the bounds of today. But research proposals of this character must not be incremental improvements of approaches that appear to be going nowhere, or into areas that seem unlikely to ever prove very important. There must always be room in a research program for the bold and visionary, but we must also consider that there are major and dangerous problems with systems that we all have literally in our hands today, and those we know with near certainty will appear tomorrow. This recommendation must be balanced by what we expect industry to address. Problems that are causing large companies to lose money are more likely to be addressed by industry than problems that do not have obvious financial implications. Problems whose solutions can lead to profitable products are likely sources for industry funding. Problems whose solutions are mandated by the laws of the United States or other large and influential nations are more likely to be addressed by industry. However, we should also remember that much industry research remains private and secret. There is value in supporting publicly available research with wide applicability, even if a few large companies might perform similar proprietary research purely for their internal benefit.

To solidify these recommendations, we recommend that the National Science Foundation prioritize research funding for privacy and security in the mobile and wireless environments in the following ways:

a) Fund projects that offer good possibilities to solve problems that have been observed in real world situations and for which good solutions are not yet known.
b) Fund projects that propose to build systems that will, at least in a proof-of-concept fashion, demonstrate such problems being directly and successfully addressed.
c) Fund projects that improve our knowledge of how people move and what computing and networking operations they perform when they move, particularly taking privacy and security issues into consideration. Many privacy and security solutions cannot be realistically tested without such knowledge, and industrial research of this kind is usually not made available to the general research community.

We also recommend that the National Science Foundation call particular attention to certain known problems in the areas of privacy and security for mobile and wireless networks. Some of these problems have proven to be very hard to solve, having already defeated early attempts or having failed to produce credible responses. Others are problems that are clearly on the horizon, and do not seem amenable to well known security techniques from other environments. These problems include:

a) Protecting a network against malicious software brought in by a mobile computer that has visited an insecure location.
b) Allowing a mobile user to gain effective control over the privacy of his movements and activities in the various places he visits.
c) Ensuring that a sensor network provides the best possible information for the longest possible period of time in situations where opponents can either disable or compromise some of its nodes.
d) Allowing a ubiquitous environment in a typical home to be sufficiently secure for normal users' purposes without requiring any but the most minimal actions on the part of such users.
e) Designing self-healing mobile and wireless network systems and mechanisms that support self-healing.
f) Finding efficient application level techniques that minimize the cryptographic overhead when the system is not under attack.
g) Protecting sensitive or classified data in mobile wireless networks operating in extreme conditions, such as disaster relief or military situations. Homeland Security requires such protection because today's terrorist is, unfortunately, a good hacker.

1.5 Conclusion

This report distills the deliberations of the mobile and wireless security experts who participated in the 2006 Workshop on Security and Privacy in Wireless and Mobile Networks (WSPWN), held in Miami, Florida in March 2006. The goal of that workshop was to offer expert guidance to the National Science Foundation on priorities in research directions in the fields of privacy and security for today and tomorrow's wireless mobile environments. The recommendations contained here come from the papers published at the workshop, the open discussions on this subject held during the workshop, and extensive discussions among workshop participants subsequent to the event.

The previous section contains many detailed technical recommendations on the areas of research we feel are likely to be most critical for the near future. In addition to these specific recommendations, the authors of this report feel compelled to point out that these areas of research are underfunded. We see regular reports of crimes and hazards related to unaddressed privacy and security vulnerabilities in today's wireless and mobile networks, and can easily foresee that the situation will only get worse as these technologies are used by more people in more situations for more purposes. Without an increase in funding in research in these areas, critical problems will remain unaddressed until they reach crisis proportions, and possibly only after a real disaster has occurred. In many of the recent stories concerning security incidents in wireless and mobile situations, there was potential for immense damage. This potential was not averted because of wonderful security technologies we have in place, but by mere chance. As it happens, it appears that the data on military flash drives sold in Afghan bazaars did not lead to US soldiers being killed in ambushes. As it happens, most thefts of laptops containing vital personal data have not lead to massive identity theft. As it happens, the worms that have already spread through wireless networks and mobility are mostly pranks or toys, not serious attempts to cause damage. But we must be aware that the possibility of true disaster was present in each of these cases. If we had done better security research in the past, we would not have had to rely on blind luck to avoid such disasters.

Part of the solution to the current vulnerabilities and dangers in the mobile and wireless world is wise choices of the research that individual researchers perform and agencies fund. However, if funding levels for this kind of research remain low, we risk having to make choices which are no more than educated guesses on where we will do research to protect ourselves and where we will leave vulnerabilities and dangers unexamined.

2 Pervasive Systems: Enhancing Trust Negotiation with Privacy Support

Jan Porekar[1], Kajetan Dolinar[1], Aleksej Jerman-Blažič[1] and Tomaž Klobučar[2]

[1] SETCCE (Security Technology Competence Centre), Jamova 39, Ljubljana, Slovenia
[2] Jožef Stefan Institute, Jamova 39, Ljubljana, Slovenia

2.1 Introduction

Pervasive or ubiquitous systems have been the subject of intense conceptual research in recent years [1,2] In favour of the sceptics, who believe that a physical world around us is complicated enough and that humankind has more important things to do than to build its digital counterpart, one can easily observe that such pervasive systems are still pure science fiction in terms of technical implementation today.

The number of electronic devices connected to the network is expected to rise exponentially and will eventually outnumber humans living on the planet. Mobile devices such as laptops, personal digital assistants and cellular phones will steadily increase in number. Standard household appliances and machines will be connected to the network and new intelligent appliances and biosensors will emerge.

The vision of pervasive systems is to integrate all those different devices in a world where computer technology will slowly disappear from everyday lives and eventually become invisible - A world in which computer systems will seamlessly adapt to user context and will help a user perform tasks by inferring his intent. A world in which a digital representation of the user, the user's data and the user's digital workplace will constantly be copied across various network nodes in order to follow the user in his real world geographical movements. Many of these devices will have a certain degree of passive and active intelligence built in and will act as sensors or reality aware processing nodes. Aside from these peripheral devices, a vast network of intelligent middleware will have to be provided in order to achieve the synchronous intelligent behaviour of the whole pervasive network.

In order for this to be achieved, a large amount of private user data, preferences, behavioural habits and other information about the user will need to be processed and exchanged among various network nodes and subsystems. With the data inferred, related conclusions will again be exchanged all over the system. In such a system, it is of paramount importance to assure privacy and maintain control of turbulent private information flow, whilst preventing leakages of sensitive private information.

Another aspect which further blurs privacy issues is diminishing of conventional role of thin, not-trusted-user-client and large-corporate-service. Pervasive systems are service oriented platforms where everything can potentially act as a service, including the user. The opposite is also true: every service will potentially be able to take on the role of a user. In pervasive systems, a user and service are simply roles that can be swapped or interchanged. These two roles merely describe the nature of the communication, since the user is the party that initiates the communication and the service is the party that replies and grants access to the user. To avoid confusion, we will use terms supplicant for the user and supplier for the service. Distributed systems are traditionally seen as environments where the user is normally not a trusted party and services are more or less trusted. In pervasive systems such as the DAIDALOS pervasive platform [9], this relation between a small user and fat service disappears or can even be intertwined.

The concepts of privacy protection are supported by three distinguishable mechanisms which conduct the process of privacy terms agreement, data access control and anonymization of the subjects involved in the process. These concepts are also known as privacy or trust negotiation, virtual identities and (access control) credentials. The first step towards protecting a user's private data is a multiparty understanding of the terms, conditions and content of private data collected and used. When a bilateral (or multilateral) agreement is reached, a selection of virtual identities is generated and activated, interpreting subjects and their context behind different levels of anonymous identifiers. The final step in the process is to relate selected identities with the user context to be used by the service and to unveil private data access control rules enforcing credentials.

The initial and principal step of privacy mechanisms is the negotiation process which defines the framework for private data protection. We therefore investigate the current state of trust or/and access control negotiation and highlight the need for it to be extended with assertions about privacy in order to satisfy the privacy constraints of the pervasive environment.

The result of such a negotiation would be: the granting of access to services and a privacy agreement that could be used by privacy enforcement systems. In the paper we also describe privacy risks of the state-of-the-art trust negotiation methods.

2.2 Trust Negotiation

Trust negotiation is a process through which mutual trust is incrementally established by the gradual exchange of digital credentials and requests for credentials among entities that may have no pre-existing knowledge of each other. Digital credentials are an electronic analogue of paper credentials used to establish trust in the every day world. Upon successful trust negotiation the supplicant is granted access to the protected resource [3,4].

During trust negotiation, the disclosure of credentials is governed by access control policies. Trust negotiation has been intensely discussed in various publications in recent years [3,4,5,6,12,13]. You will also find a brief description of a trust negotiation protocol in this document.

The parties involved in trust negotiation will be named the supplicant and the supplier. The supplicant is the party that requests access to resource R, and the supplier is the service providing it. Trust negotiation protocol consists of two types of messages which are exchanged between the supplicant and supplier:

1. Requests for credentials or resources;
2. Disclosures of credentials or resources.

In the text below we describe a typical negotiation example. In the first step of negotiation a supplicant sends a request to a supplier for access to the resource R. The supplier can either grant access to the resource R directly or request an additional set of credentials C1 to be sent first. In this case, the supplicant can decide whether he trusts the supplier enough to disclose C1. If the supplicant doubts about the supplier's trustworthiness, he can reply by requesting an additional set of credentials C2 from the supplier. When the supplier replies by presenting credentials C2, the supplicant replies by sending credentials C1 back to the supplier. Because all requests have been satisfied and appropriate credentials presented by both parties, the supplicant is granted access to the requested resource R. For better clarity, the example is presented in Fig 2.1.

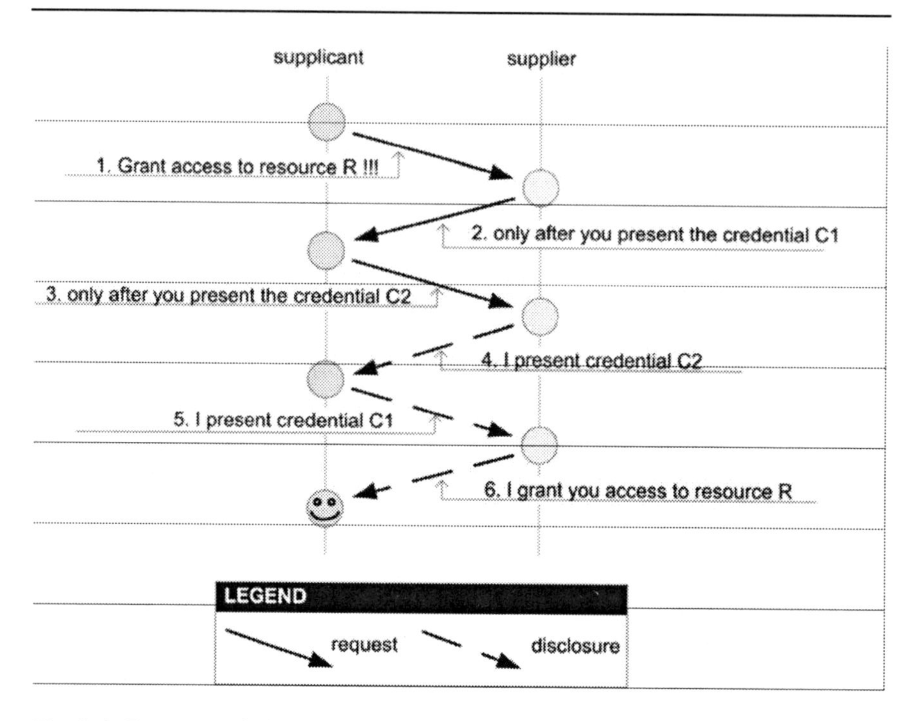

Fig. 2.1. Trust negotiation schema

In general, negotiation may consist of several steps. In each step, one of the two parties may disclose some credentials that were requested by the other party during the previous step. In addition to the disclosure of credentials a party may choose to request additional credentials to be disclosed by the other negotiating party, before it trusts the other party enough for the requested credential to be revealed. The exact flow of the exchanged credentials depends on decisions made by each party involved in negotiation and is referred to as "strategy" [4,6]. Strategies determine which credentials are to be revealed, at what times and when to terminate the negotiation. Strategies can be either more liberal or more conservative in terms of willingness to disclose the information. In this manner the trust is gradually established between both negotiating parties.

2.3 Weaknesses of Trust Negotiation

We define privacy risk, or privacy threat, as a measure of the possibility that private data, which is desired to stay private, is revealed without the owner having the ability to prevent this. A Privacy leak is defined as any unintentional disclosure of private data, either as a consequence of negligence, weak privacy provision methods, or capability to compromise these. Thus, any leak is also a threat, fulfilled threat, and it depends on degree of information leaked how big threat it is.

The main goal of the trust negotiation process described above is to grant the supplicant access to the requested resource. The very fact that sensitive attributes are revealed during the negotiation process calls for attention, in fact under certain conditions even access control policies can be regarded as private or sensitive information that needs to be handled with special care.

Apart from the straightforward disclosure of private information during manipulation, privacy can be at risk in a far more indirect and opaque sense. Pervasive environments make information processing highly intensive and penetrating and can render small pieces of information which can be stepping stones to the disclosure of greater secrets. Quite naturally, a large amount of personal information will already be available to systems in the pervasive environment after a longer period of use of the system. Although data have probably been made adequately anonymous as far as possible (compare methods for pseudonymizing in [7] or the virtual identity approach in [9]), inference capabilities of a pervasive environment can aid in correlating sets of anonymous data with each other. This can make aggregating correlated data possible and resolving personal profiles to an extent where it is finally unambiguous in relation to one unique person. This possibility is called linkability of (anonymous) personal information. We want to avoid this is the effect by all means and aggravating this is one of the major concerns of identity management systems in a pervasive environment (compare again [7, 9]). For this reason we compare the pervasive environment to the example of a chaotic dynamic system with respect to the degree and significance of information disclosed over time. Any information available can consequently result in a disclosure of certain private data which was not intended in the first place – thereby resulting in a privacy leak. The measures taken to prevent linkability can therefore never be exaggerated and every procedure involved in disclosing private data has to be evaluated from this viewpoint.

In this section we study weaknesses of the described trust negotiation methods that can lead to privacy leaks in the sense of the straightforward disclosure of private data, for example disclosing a sensitive credential, or due to linkability. Some of the weaknesses have already been discussed in literature [4] and some of them reflect our original work. The related leaks and threats pertain to supplier as well as to supplicant, especially straightforward disclosure. But while the supplier is often (but not necessary) a publicly known entity, it is characteristic for the supplicant to focus more relative importance on maintaining anonymity and thus linking is of more threat to supplicant.

Disclosing credentials could be a privacy risk. When the supplicant is requested to disclose certain credentials during the negotiation, it may react to the request in various ways. If the credential is not valuable enough to the supplicant in the context of the current negotiation, the supplicant may choose to willingly present the credential without much hassle. An example of such a negotiation situation would the case where a supplicant is trying to buy a camera from an online store and he gets offered a discount if he is willing to present credentials that prove that they are a citizen of the European Union. If user is not concerned with anyone finding out that he or she is indeed a citizen of EU, disclosing the credential results in minimal privacy threat. On the other hand if a British Secret Service agent is asked to provide an MI5 membership credential in order to get discount on a camera he is trying to buy, it is a obviously a different matter. MI5 membership credentials is sensitive information that is not to be shown to just anyone and disclosing it could be a serious privacy risk, thus highlighting another category of linking private data.

Obviously a disclosure of credentials is a potential privacy leak. But the answer to the request for certain credentials can also potentially yield information. An example of such an information leak would be that of a supplier requesting a supplicant present an “MI5 Membership Credential”. In order for the supplicant to determine if the supplier is trusted enough, the supplicant asks the supplier to provide the “Ring of Secret Service trusted Membership” credential. When the supplier receives the additional request from the supplicant it can assume with a certain degree of probability that the supplicant possesses the credential that was requested in the first place. The amount of probability depends on different negotiation strategies that supplicant chooses to pursue and his ability to bluff.

Not disclosing credentials could in some cases also yield useful information for linking. The sole fact that the supplicant has attempted to access a supplier resource could limit the scope of possible supplicants. Credentials may indicate that the supplicant belongs to one of two mutually disclosing classes of supplicants. Inability to provide the requested credential, either due to disagreement or failing to posses one, could also enable the supplier to categorise the supplicant and thus to help linking of data in the future.

Disclosing access control policies could be a privacy risk. When a supplier is asked to grant access to the requested resource it can provide the feedback about requested credentials back to the supplicant in many different ways. If the supplier has its access control policies on public display, it is fully acceptable for it to return the whole policy back to the supplicant. Afterwards the supplicant accepts can then navigate through many parallel options in order to find the combinations of credential disclosures that are optimal for him. While this is fully acceptable if the supplier is a governmental organisation that provides its services to citizens and has published access control policies; it is not the case when a supplier is a service providing sensitive resources. For example if a supplier is a server of the British Secret Service, which is providing sensitive top-secret data to its agents on the road it will not publish its policies to the public, since the policies contain valuable data on the organisational hierarchy of the supplier, and revealing the policies would provide valuable information which could be potentially misused. Instead, the supplier will try to minimize the amount of information provided at each step of negotiation by requesting one credential after the other or maybe choosing not to provide information detailing which credentials should be disclosed to the user at all.

Exploiting negotiation to steal private data – trust negotiation piracy. With careful design of trust negotiation algorithms it can be possible to exploit the trust negotiation protocol to serve private information under pretext of a legal purport. The purport is more likely to be abused by a supplier role in the context of a service provider with a range of services, promised large enough to relate to a wide scope of interesting categories about supplicants. Consider following example.

The supplier is a service offering bets in several categories, depending on the supplicant profile. The supplicant is provided a possibility to apply for the service as a pseudonymous user with its true identity hidden. Systems for auditing in a pervasive platform architecture make non-repudiation of debts possible (compare [10] for example). Although the service might actually provide what it has claimed to provide (it has also been certified so), let us suppose that it also has the intention to aggregate the profile information of supplicants in order to (at least partially) determine their identity. The handshaking could possibly proceed as follows:

1. Supplicant: accesses the service web portal.
2. Supplier: "We offer several categories for bets: bets on the outcome of sport events, bets on the outcome of political events, bets on the results of science research ... Select your interest ..."
3. Supplicant: chooses politics.

4. Supplier: "Which event from following: the outcome of upcoming elections, ..., the outcome of the acceptance of last week's formal proposal for amendment to act 26.8/2005, ..."
5. Supplicant: chooses an event.
6. Supplier: demands a credential that supplicant's age is above 18.
7. Supplicant: demands credential that supplier will not use this information for any other purpose than service provisioning.
8. Supplier: provides the credential.
9. Supplicant: provides the credential.
10. Supplier: "We only allow bets above 1.000,00 € for this category." Demands a credential on supplicant's financial liability.
11. Supplicant: demands credential that supplier will not use this information for any other purpose than service provisioning. Supplicant: provides the credential.
12. Supplier: provides the credential.
13. Supplicant: provides the credential.
14. Supplier: demands a credential that supplicant is not employed in a state department service. The supplier imposes the restriction based on the fact that access to privileged information would help to win bets, and is not allowed.
15. Supplicant: withdraws.

If we analyze the above sequence we can figure out that supplier could deliberately design categories to address classes of people and their interest. When the supplicant has revealed his interest via selection in step 3, the supplier can then assign the supplicant to this category. Further suppose that the supplicant designed events according to increasing political awareness, as carefully as it can imply certain political skills and positions. Then selection under step 5 further scopes the category.

After step 5 the true exchange of credentials in the sense of trust negotiation starts. The resource here negotiated for is a betting account on a respective event. After each credential is received, the supplicant can determine a more focused scope of potential persons satisfying specific attributes: age, financial profile and associated implications ... And finally, the supplicant can also determine why a supplicant has withdrawn – possible causes could involve people with significant political positions. Moreover, the sequence could be designed as to gradually lead the supplicant through the disclosure of credentials with less privacy threat, and then to present requests for credentials with higher threat so that many credentials will have already been disclosed before the supplicant finally refuses to make further disclosures and withdraws.

Similar services already exist in today's Internet world and there is no reason to think that such scenarios would not appear in a pervasive environment. The supplier could have sophisticated systems for reasoning in place, as this is not unusual aspect of pervasive system capabilities. If we assume an appropriate degree of information processing and a large enough period of time, the supplier can deduce information about people concerning their bets, their financial status, and their interests – and can enable the linking of this information to real persons and then use this for blackmailing and other illegal activities. With this in mind, the above resolutions are not really unbelievable.

The first weakness of trust negotiation apparent from the above example is that disclosing interest in step 3 and 5 is not included in trust negotiation. If we consider that in pervasive systems it will be practically impossible for a supplicant to perform or even only supervise privacy related procedures because of the high degree of information exchanged in very short time periods, trust negotiation and the remaining subsequent enforcement has to be done in a computer aided manner. The supplicant will rely on the privacy subsystem in order to have privacy adequately maintained. Disclosure of this kind of information as in steps 3 and 5 was done willingly, but supplicant software components were not given the chance to evaluate the consequences and make this subject to identity management. Thus this could represent a privacy threat and allow future privacy leaks. General terms about the attitude towards abstract notions of disclosing, as for example a specific interest, which needs to be identified in the overall negotiation and provided for processing to enforcement systems. For example, this is necessary for identity management if it should be able to extract information on how big a threat of linking is with respect to the disclosed interest and what virtual (or partial) identity should be selected.

The second weakness is that at the end of the above sequence the supplicant didn't get access to the resource, but has still revealed quite a large amount of personal information. Trust negotiation cannot happen in pure general terms arguing on meaning of resources and credentials in advance. By applying purely general terms of negotiation we could resolve collisions in attitudes of supplier and supplicant before any resources or credentials are disclosed, and thus supplier is left only information about supplicant attitudes, while credentials were preserved.

2.4 Extending Trust Negotiation to Support Privacy

In the document above we have shown the need for current trust negotiation to be extended to support privacy issues. Generally speaking two different approaches could be undertaken to achieve this.

The first one is to introduce negotiation of general terms of privacy practices exercised on information both parties are about to disclose in the future and do this before trust negotiation. We have chosen to name this new kind of negotiation a privacy negotiation. Here we keep this separated from trust negotiation. To facilitate such a negotiation no resource is explicitly necessary to be disclosed in order to achieve a resulting agreement; instead we argue about the attitude towards opposite side practices with respect to manipulating private data. A way of formal description of resources is required, and related semantics and a means of semantic processing so that reasoning on relevant statements can be performed. For a possible example of a suitable framework see [14]. Privacy policies need to be specified in a formal way suitable for computer processing. Privacy policies need to be specified in a formal way suitable for computer processing. Much interesting research for this has been done with respect to an ontology approach; compare for example [16]. We will avoid presenting here detailed techniques to technically facilitate such a formal negotiation as the scope of this paper focuses mostly on specific problems of protocols and related threats. The outcome of a negotiation is a set of statements expressing the attitude of a supplier and supplicant to the matters exposed in the negotiation, whose meaning can be resolved against resources. This set is respected as a privacy agreement, a formal document which is mutually signed. After this negotiation the parties would proceed and start a well known trust negotiation.

The second approach extends current models of trust negotiation to support privacy negotiation requests and corresponding privacy negotiation agreements as responses, relying on the approach from the previous paragraph. After successful negotiation all privacy agreements from various levels of negotiation are merged into a final privacy agreement, while trust negotiation itself is still performed in parallel.

In a naive way the first approach could be implemented using existing solutions. For the privacy negotiation practice, P3P policies can be used [15]. The user is presented the P3P privacy policy when trying to use the service. The only option for the user is to accept the privacy policy presented by the service and opt in, or out, of certain issues. Beside the mentioned opting not much of negotiation takes place using P3P policies. With user accepting the P3P terms of the privacy policy privacy agreement is reached. The next stage would be to negotiate for access to requested resource using one of the negotiation systems available today (i.e. Peer Trust, Trust Builder, etc.) (see [12,13]). The problem with this approach is that in many cases trust cannot be evaluated solely on a general basis but some credentials have to be disclosed in order to proceed. There are several reasons why pure privacy negotiation cannot efficiently bring the negotiation to an end. Negotiating general terms would result in resolving a very huge problem space of possible solutions to the negotiation because a peer (supplier or supplicant) doesn't have options clearly defined; a peer explicitly requests a credential in order to continue negotiation; etc. This leaves us with no other option than merging privacy negotiation and trust negotiation into a common framework.

2.5 Proposed Trust Protocol Extended to Support Privacy

Based on the statements in the previous section we construct a protocol supporting integration of privacy measures into trust negotiation. Four different types of assertions are part of the protocol:

1. request for credentials or resources
2. disclosure of credentials or resources
3. request to agree with certain privacy practices (proposals of privacy agreements)
4. acceptance of privacy practices proposals (accepted and signed privacy agreements)

The parties involved in a process of negotiation are a supplicant and a supplier. An example of negotiation is described below that corresponds to Fig 2.2. The supplicant is the party requesting access to a specified resource R and the supplier is the service providing this resource. In the first step of negotiation the supplicant sends a request to access R to the supplier. The supplier can either grant access to the supplicant or request additional credentials C1 to be revealed. In case of additional credentials being requested the supplicant can either disclose the requested credential or reply back to the supplier with another request.

But, as a difference to an ordinary trust negotiation, it is now possible to follow data minimisation principles (for definition see [8]): we don't want to disclose the requested credential at this point as we're not sure whether negotiation will succeed at all. In case the negotiation was unsuccessful, we would end with a series of credentials disclosed, but no real effect achieved (as described in

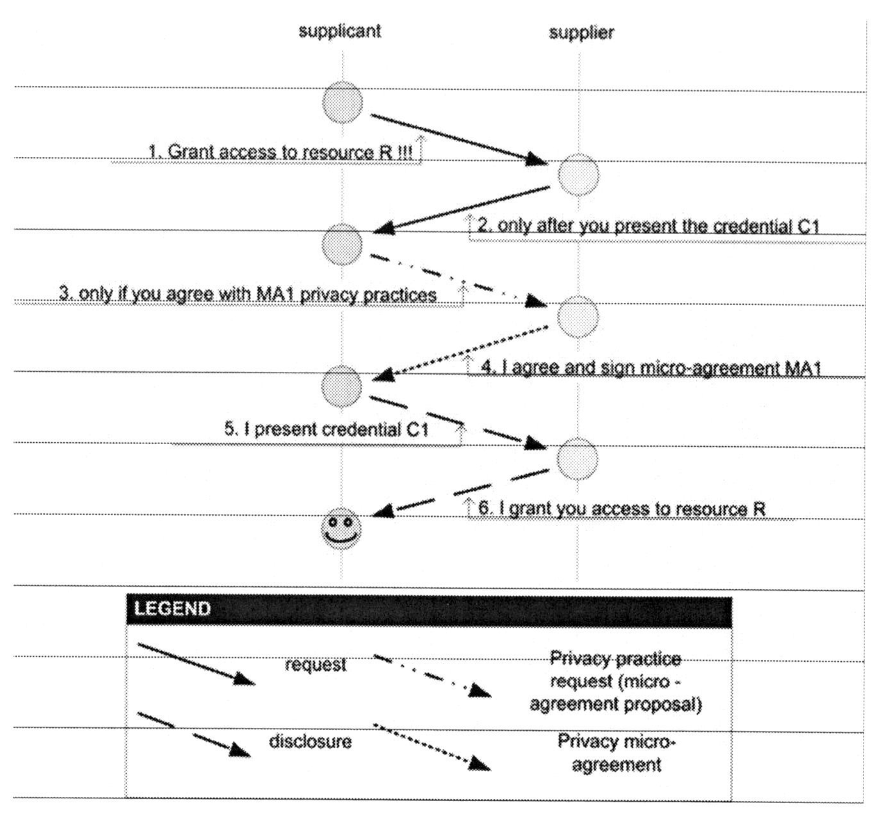

Fig. 2.2. Schema of privacy extended trust negotiation

Chapter 3). From the data minimisation principle aspect this is not allowed. Data minimisation principle imposes a requirement for amount of private data disclosed for service provisioning being as small as possible, disclosing only really necessary information. But in this case we have possibly already disclosed a significant amount of private information before negotiation failed by revealing credentials about various attributes associated to user's private life. Instead of this here we rather argue about privacy terms in general at this point, applying only privacy negotiation until this is still possible from logical viewpoint.

The partial agreement that was done in sense of privacy negotiation sequence will from now on be called a micro-agreement to avoid confusion with a cumulative privacy negotiation agreement that aggregates all the micro-agreements which were reached and signed during the process of privacy negotiation. This cumulative privacy negotiation agreement is mutually signed as well.

2.6 Privacy Agreement

The privacy negotiation agreement consists of many independent micro-agreements (MA). Each of the micro-agreements being mutually signed by both parties involved in negotiation in order to limit potential repudiation of the agreement.

No matter of the result of negotiation the micro-agreements are bound into privacy negotiation agreements after the negotiation is finished. If negotiation outcome was successful, the privacy negotiation agreements are mutually signed by both parties. In case the negotiation was terminated before access control was granted, the micro-agreements can still be bundled into a privacy negotiation agreement. This way potential misuse of information about sensitive attributes is prevented in at least a formal juridical way.

Privacy agreement can be viewed as a digital analogue of the paper based contracts and agreements exchanged by parties every day, which consist of obligations that both parties involved in a contract or agreement need to fulfil. In real world examples these obligations are usually payment on one hand and providing resources, products or services on the other. In the context of privacy agreements the obligations are private or sensitive information on one hand and privacy practices on the other. By the term privacy practices we refer to the way private information is handled, to which 3rd parties it will be transferred and how it is inferred, aggregated or statistically manipulated.

Privacy agreement is a starting point for different privacy enforcement systems to act upon. These systems can either be identity management components or components that are analogous to legal prosecution systems of real world, such as auditing and logging components in DAIDALOS [9]. The agreements are taken as input information for systems determining whether the services or users comply with promised privacy practices.

If one of the parties denies signing the privacy negotiation agreement when negotiation was not successful and resulted in termination, it can be treated as intent of privacy agreement misuse and this can immediately be reported to privacy enforcement components of the system.

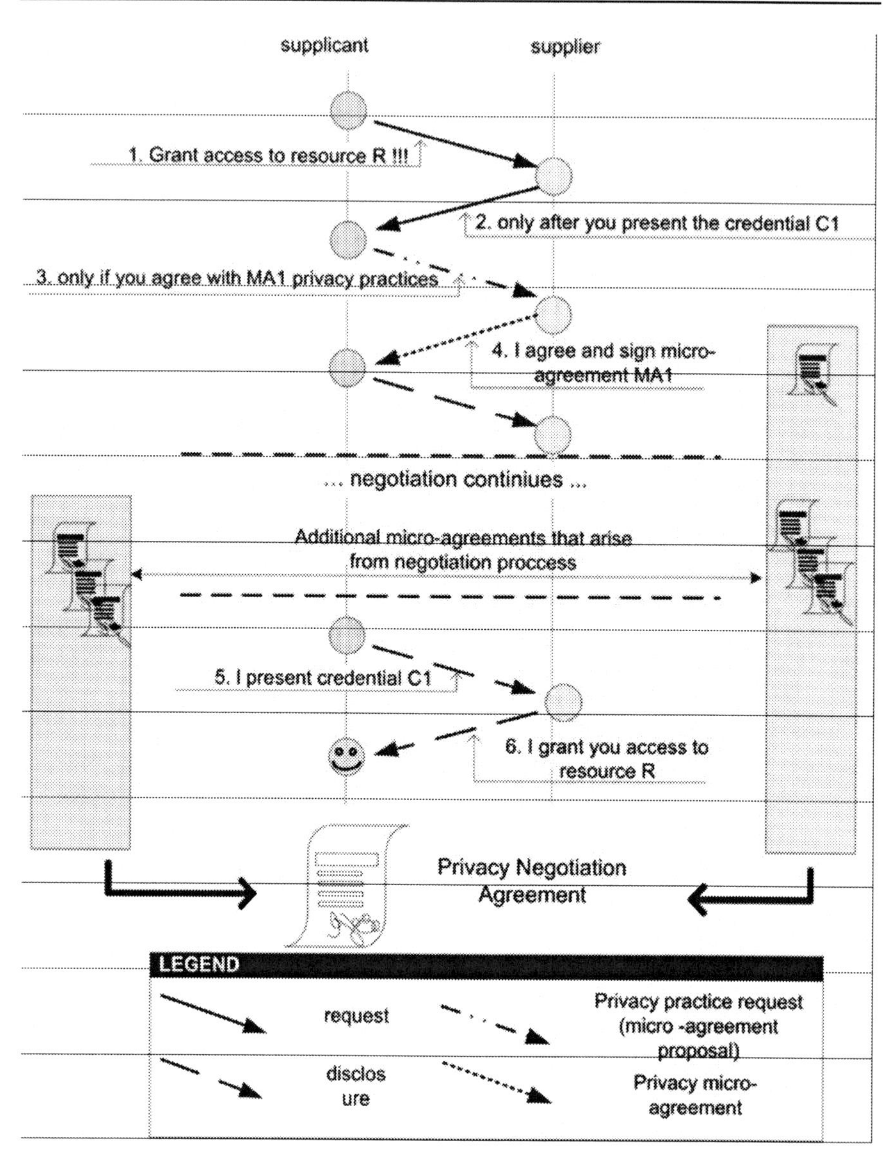

Fig. 2.3. Privacy Negotiation Agreement is an aggregation of micro-agreements

2.7 Conclusions

The privacy policy negotiation process involves gradual step-by-step disclosure of attribute values between both the supplier and the supplicant and is therefore a possible source of privacy leakage. Both supplier and supplicant need to negotiate firmly and conservatively in order to minimize this leakage. If a conservative strategy is used consistently, less and less negotiations will end in a positive resolution. In the current model there is no way for the user to determine the type of negotiation strategy to use with the given service – whether the user initially should have conservative or liberal stance towards the service.

In order to expand this, the current privacy negotiation models should be composed with existing trust modelling techniques using the trust and risk computation modelling techniques. Fusion of these trust management systems, privacy negotiation and identity management models should introduce a concept of initial measure of trust between user and service. Upon this trust the negotiation strategy could be chosen (either conservative – privacy paranoid, neutral, or liberal – give all information away like). This trust would be constantly updated through a loop – like feedback of trust reporting. The initial measure of user's trust is based on the aggregation of previous experience of users with the service using different trust and risk computation techniques [11].

References

1. Saha D, Mukherjee A (2003), Pervasive Computing: A Paradigm for 21st century. IEEE Computer Society.
2. Satyanarayanan M (2001) Pervasive computing: Vision and Challenges. IEEE Personal Communications, IEEE Computer Society
3. Bhargava B, Lilien L, Rosenthal A, Winslett M (2004) The Pudding of Trust. IEEE Intelligent Systems, IEEE Computer Society
4. Seamons KE, Winslett M, Yu T, Yu L, Jarvis R (2002) Protecting Privacy During On-line Trust Negotiation. Lecture Notes in Computer Science, Springer-Verlag GmbH , Volume 2482 / 2003, pp. 129–143.
5. Seamons KE, Winslett M, Yu T. (2001) Limiting the Disclosure of Access Control Policies during Automated Trust Negotiation. Proc. symposium on network and distributed systems security, NDSS
6. Chen W, Clarke L, Kurose J, Towsley D (2004) Optimizing Cost-sensitive Trust-negotiation Protocols. Technical Report 04-29, Dept. of Computer Science, UMass, Amherst
7. Prime Consortium (2004) PRIME – Architecture version 0. Deliverable D14.2.a
8. Prime Consortium (2005) PRIME – Framework version 1. Deliverable D14.1.a
9. DAIDALOS Consortium (2004) DAIDALOS pervasive systems privacy and security framework and mechanisms, Deliverable D421

10. DAIDALOS Consortium (2004) A4C Framework Design Specification. Deliverable D341
11. Richardson M, Agrawal R, Domingos P (2003) Trust Management for the Semantic Web. Proc. 2nd International Semantic Web Conf., LNCS 2870, Springer-Verlag, pp. 351–368.
12. Nejdl W, Olmedilla D, Winslett M. (2004) PeerTrust: Automated Trust Negotiation for Peers on the Semantic Web. Secure Data Management, pp. 118–132.
13. Winslett M, Yu T, Seamons KE, Hess A, Jacobson J, Jarvis R, Smith B, Yu L (2002) Negotiating trust in the Web. Internet Computing, IEEE, Nov/Dec, Vol. 6, pp. 30–37.
14. OpenCyc, http://www.opencyc.org.
15. Wenning R (2005), The Platform for Privacy Preferences 1.1 (P3P1.1) Specification. W3C Working Draft
16. Nejdl W, Olmedilla D, Winslett M, Zhang CC (2005) Ontology-Based Policy Specification and Management. Proceedings of European Semantic Web Conference (ESWC2005), May/Jun, Heraklion, Greece.
17. Porekar J, Dolinar K, Jerman Blažič B (2007) Middleware for Privacy Protection of Ambient Intelligence and Pervasive Systems. WSEAS Transactions on Information Science and Applications, Issue 3, vol 4, March, p/pp 633–639

3 Applying Trust in Mobile and Wireless Networks

Dagmara Spiewak and Thomas Engel

SECAN-Lab
University of Luxembourg, 6, r. Richard Coudenhove-Kalergi,
L-1359 Luxembourg

3.1 Introduction

Security-sensitive data and applications transmitted within mobile ad-hoc networks require a high degree of security. Because of the absence of fixed base stations and infrastructure services like routing, naming and certification authorities, mobile ad-hoc networks differ highly from traditional wireless networks. In MANETs, nodes may join and leave the network arbitrarily, sometimes even without leaving a trace and the network topology may change dynamically. Consequently, it is very important to provide security services such as authentication, confidentiality, access control, non-repudiation, availability and integrity. Due to the fact that central trusted third parties (TTP) are not appropriate in mobile ad-hoc network settings, the notion of *Trust* becomes more and more important. Although *Trust* is well known in everybody's life, the formal definition poses several challenges. So far, subjective interpretations and notions about the word *Trust* lead to ambiguousness of the term. In [19] Pradip Lamsal presents a wide expertise on the description of trust in networks and its relationship towards *Security*. Nowadays, the concept of *Trust* in the computing environment mainly appears in combination with e-commerce on the Internet, for example in the *PayPal* Payment System used for securely transferring money over the Internet. In [10] a direct comparison between *Trust* systems applied in the Internet and the requirements for *Trust* systems in spontaneously emerged mobile ad-hoc networks, where the *Trust* establishment has to be performed without the presence of a *Trust* infrastructure, is presented. Due to the dynamic character and quick topology changes, *Trust* establishment in mobile ad-hoc networks should support among others a short, fast, online, flexible, uncertain and incomplete *Trust* evidence model and should be independent of pre-established infrastructures. In this context, Pirzada and McDonald [21] emphasize the interdependency of *Trust* and security, while security is highly dependent on trusted key exchange and trusted key exchange on the other side can only proceed with required security services. Moreover, ad-hoc networks rest on trust-relationships between the neighboring nodes that evolve and elapse on the fly and have typically only short durability. Assuming such an environment misleadingly as cooperative by default would ignore the high vulnerability to attacks on these trust relationships. Particularly selfish, malicious, or faulty nodes pose a threat to availability and functionality of mobile ad-hoc networks and may even exploit these trust relationships in order to reach desired goals. To overcome these difficulties, again *Trust* in mobile ad-hoc networks has been used, introducing several conditions, such as the presence of a central authority. Unfortunately, these solutions are mainly against the real nature of spontaneous mobile ad-hoc networks. The concept of *Trust* Management is defined by Audun Jøsang, Claudia Keser and Theo Dimitrikos in [18] as "The activity of creating systems and methods that allow relying parties to make assessments and decisions regarding the dependability of potential transaction involving risk, and that also allow players and system owners to increase and correctly represent the reliability of themselves and their systems".

The following section presents feasible attacks in mobile ad-hoc network settings, prior to the descriptions of different *Trust* Models in the subsequent sections.

3.2 Attack Analysis for MANETs

Two different kinds of security attacks can be launched against mobile ad-hoc networks, *passive* and *active* attacks. The attacker rests unnoticed in the background while performing a passive attack. He does not disturb the functions of the routing protocol, but he is able to eavesdrop on the routing traffic in order to extract worthwhile information about the participating nodes. Running an active attack, the attacking node has to invest some of its energy to launch this attack. In active attacks, malicious nodes can disturb the correct functionality of the routing protocol by modifying routing information, by redirecting network traffic, or launching Denial of Service attacks (DoS) by altering control message fields or by forwarding routing messages with falsified values. Attack categories that can occur associated with vulnerabilities of mobile ad-hoc systems are described below.

3.2.1 Passive attacks

A malicious node in the mobile ad-hoc network executes a passive attack, without actively initiating malicious actions. However, he can fool other network participants, simply by ignoring operations. Furthermore, the malicious node attempts to learn important information from the system by monitoring and listening on the communication between parties within the mobile ad-hoc network. For instance, if the malicious node observes that the connection to a certain node is requested more frequently than to other nodes, the passive attacker would be able to recognize, that this node is crucial for special functionalities within the MANET, like for example routing.

Switching its role from passive to active, the attacker at this moment has the ability to put a certain node out of operation, for example by performing a Denial of Service attack, in order to collapse parts or even the complete MANET. Additional examples of a passive attack represent selfish nodes. They derivate from the usual routing protocol for the reason of preventing power loss for instance by not forwarding incoming messages. In [5] the importance of *Trust* is emphasized in order to isolate these malicious nodes and to be able to establish reputation systems in all nodes that enable them to detect misbehavior of network participants.

3.2.2 Active attacks

Active attacks mainly occur subsequent to passive attacks, for example after the malicious node finished eavesdropping required information on the network traffic. The variety of active attacks on mobile ad-hoc networks is similar to the attacks in traditional and hierarchical networks. But due to the lack in infrastructure and the vulnerability of wireless links, the currently admitted routing protocols for mobile ad-hoc networks allow launching also new types of attacks. Compared to passive attacks, malicious nodes running an active attack can interrupt the accurate execution of a routing protocol by modifying routing data, by fabricating false routing information or by impersonating other nodes. Basically, active security attacks against ad-hoc routing protocols can be classified in three groups [23], such as *integrity*, *masquerade* and *tampering attacks*.

Integrity Attacks in MANETs

Particularly attacks using *modifications* are aimed against the integrity of routing information. By launching this type of attack, the malicious entity can drop messages, redirect traffic to a different destination, or compute longer routes to the destination in order to increase the communication delays. For example, by sending fake routing packets to other nodes, all traffic can be redirected to the attacker or another compromised node. An example of a *modification* attack is the set-up of a *Blackhole* [22]. First of all, the malicious node analyzes the routing protocol by the use of a passive attack, like eavesdropping information on the network traffic. Subsequently, this node lies and announces itself, during the route discovery phase of a routing protocol, as knowing an accurate path to the requested target node, in order to be able to intercept packets. Finally, all packets are transferred to the attacker's node and he discards all of them. Consequently, the malicious node, which is controlled by the attacker, represents the *Blackhole* in the MANET, where all packets will be swallowed.

As an extension of the *Blackhole* attack, the active attacker might generate a *Greyhole* [11]. In this case, the malicious *grey* node has the ability to switch its course of action from forwarding routing packets or discarding others. The decisions of its behavior depend on the intention of the attack. For example, for the purpose of isolating particular nodes in the MANET the malicious *grey* node drops packets which pilot towards their destination. Packets meant for other nodes rest unmodified und are forwarded to their destination accordingly.

Even trickier is the generation of a tunnel in the network between two or more cooperating and by the attacker compromised malicious nodes that are linked through a private network connection within the MANET. This attack is known as a *Wormhole* [12]. It allows the attacker to short-cut the normal flow of routing messages by the construction of a fictitious vertex cut in the network that is controlled by the two cooperating malicious nodes. The attacker records packets or parts of packets at one selected location in the MANET. After tunneling them to another point in the MANET, the attacker replays the packets into the network. In particular, ad-hoc network routing protocols are vulnerable to *Wormhole* attacks. For example, launching this attack against a routing protocol allows the attacker to tunnel each ROUTE REQUEST packet, which is transmitted during the route discovery phase, straight to the target destination node. Consequently, any routes other than through the *Wormhole* are avoided from being discovered. By this technique the attacker has the capability to create an appearance to know the shortest path to a desired destination node. This grants the attacker an exceptionally high probability of being selected by the routing protocol to forward packets. Once selected, the attacker is able to subsequently launch a *Blackhole* or *Greyhole* attack by discarding selected packets.

Furthermore, *Wormhole* attacks empower the attacker to influence the neighbor discovery functionality of several routing protocols. For example, assuming node A wishes to communicate with its neighbors and tries to knock at their doors by sending a HELLO broadcast packet. At the same time the attacker uses the *Wormhole* to tunnel this packet directly to node B.

On the other side he tunnels all HELLO packets sent by B directly to node A. Finally, A and B belief that they are neighbors, which would cause the routing protocol to fail to discover routes when they are not really neighbors. Additional advantages of the *Wormhole* for the attacker are his possibility to discard selected data packets or to maintain a Denial of Service attack, because no other route to the destination can be determined as long as the attacker controls the *Wormhole.* Yin-Chun Hu, Adrian Perrig and David B. Johnson introduce in [12] a mechanism, called *"Packet Leashes"* for effectively detecting and defending against *Wormhole* attacks by limiting the transmission distance of a link. The authors present the *TIK* protocol which implements temporal leashes using hash trees.

Both, *Blackhole* and *Wormhole* attacks belong to the group of *Byzantine Attacks* in ad-hoc networks and are discussed in [2]. The scheme of *Wormhole* can be even extended to the concept of *Byzantine Wormhole* attacks. The difference to traditional *Wormhole* attacks is the fact that in traditional *Wormhole* attacks the attacker can fool two honest nodes into believing that there exists a direct link between them. But in the *Byzantine* case the *Wormhole* link exists between the compromised nodes and not between the honest nodes, which means that the end nodes cannot be trusted to follow the protocol accordingly.

Therefore, the previously mentioned *"Packet Leashes"* [12] are effective against traditional *Wormhole* attacks but they can not be used to discover and to prevent the extended *Byzantine Wormhole* attacks. Figure 3.1 shows the classification of these attacks in MANETs.

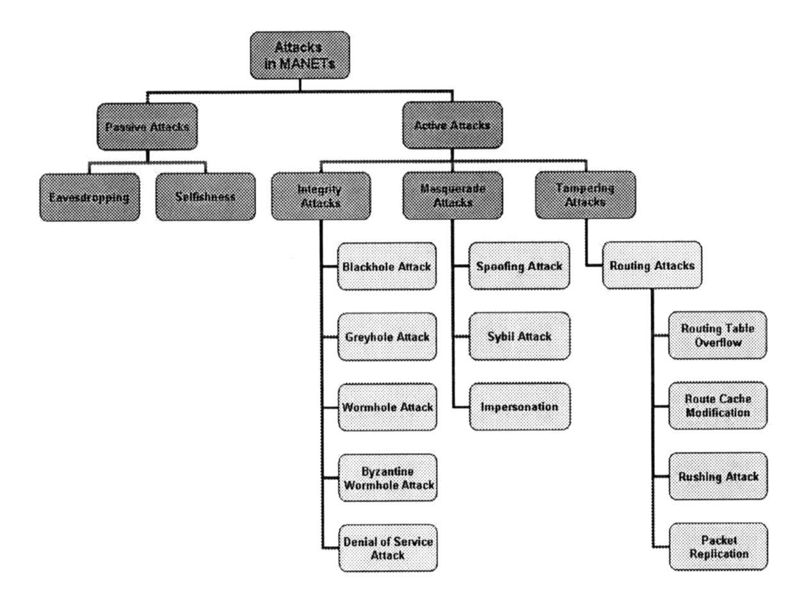

Fig. 3.1. Classification of Attacks in MANETs

Masquerade Attacks in MANETs

By masquerading as another node, malicious nodes can run many attacks in a network. These types of attack are often known as *Spoofing*. The attacker modifies either the MAC or the IP address in outgoing packets in order to adopt another identity in the network and appear as a good-natured node. By this technique he is then able to operate as a trustworthy node and can for example advertise incorrect routing information to other participants of the network. Creation of loops in the routing computation is one famous example of this exploit and results in unreachable nodes or a partitioned network. Another dangerous attack in MANETs is known as the *Sybil Attack* [7]. Here malicious nodes may not only impersonate one node but can even represent multiple identities by maintaining false identities. This attack particularly weakens systems and protocols that employ redundancy. Basically, redundancy is deployed to resist security threats from faulty or malicious network participants and is mostly used to ensure that transmitted packets are forwarded from node A to node B accordingly. By launching a *Sybil Attack* the attacker can pretend that the supposedly different paths are formed by disjoint nodes, although in reality these paths share at least one node which is the attacker's one.

Particularly MANETs that apply a Recommendations-Based Trust Model are vulnerable to *Sybil* attacks. Here the malicious node, which represents multiple identities, can generate fake recommendations about the trustworthiness of a particular node in order to attract more network traffic to it. This offers the attacker an ideal starting point for subsequent attacks, like for example the *Byzantine Wormhole* attack. Generally, forging of multiple identities for malicious intent leads to a set of faulty nodes in the network which results in compromising of all reliability-based network models.

Tampering Attacks in MANETs

This group of attacks, often called *Fabrication Attacks,* is based on the generation of falsified routing messages. Because of the fact that these routing packets are received as valid, fabrication attacks are very difficult to identify and trace. An example for such an attack is the in [13] introduced *Rushing Attack* which acts as an effective Denial of Service attack against all currently proposed on-demand ad-hoc network routing protocols, including those designed to be secure. Launching this attack, an attacker rapidly spreads routing messages all through the network, disabling authorized routing messages with the consequence that other nodes delete them as multiple copies. Obviously, also computational routes to a destination can be canceled by constructing routing error messages, asserting that the neighbor can not be reached. For this reason, since flooding is the famous mechanism used by on-demand routing protocols to establish paths, disturbing flooding is an effective attack against these kinds of protocols.

Considering the routing strategy of an on-demand ad-hoc network protocol, where node A wishes to obtain a route to a destination node B. Node A floods the mobile ad-hoc network with ROUTE REQUEST packets. In order to limit the network traffic, each intermediate node C forwards only one ROUTE REQUEST packet from any Route Discovery phase or even only the ROUTE REQUEST packet reaching C at first will be forwarded by C. If the attacker launches falsified ROUTE DISCOVERY sessions for non-existing destination nodes and if the attacker's ROUTE REQUEST packet reaches the intermediate node C prior to the ROUTE REQUEST packet from node A, then the legitimate REQUEST will be discarded by C and the attacker's REQUEST will be forwarded accordingly. With this technique the attacker is able to isolate certain nodes in the MANET or can even partition the network. Otherwise, if the attacker's rushed ROUTE REQUEST packets are the first to reach every neighbor of the target node B, then any route discovered by this ROUTE DISCOVERY process will include a hop through the attacker. Hence, node A will be unable to discover any trusted route, without the attacker's influence, to the target node B. In order to speed-up the broadcast of falsified ROUTE REQUEST packets the attacker can combine the *Rushing* attack with the *Byzantine Wormhole* attack to create a tunnel for his ROUTE REQUEST packets.

Actually, the fact that only the first ROUTE REQUEST packet is forwarded by an intermediate node C is not necessary for the attacker to be able to launch this kind of attack. The *Rushing Attack* can be extended to compromise the functionality of any protocol that forwards any particular ROUTE REQUEST packet for each ROUTE DISCOVERY process.

3.3 Existing Trust Models

The establishment of *Trust* as a component of security services in networks or as a foundation for succeeding security tasks resounds throughout the land. However, many solutions misleadingly introduce *Trust* as a matter of course but simultaneously using it as the basis for further security issues, such as for creating confidentiality, integrity, authentication or non-repudiation, without even constructing a conclusive *Trust* metric. This section presents already existing trust models, with the aim to expose their differences, before starting to examine novel models in the subsequent section. As in [25] clarified, *"trust is interpreted as a relation among entities that participate in various protocols"*. The trustworthiness of a certain entity depends on the former behavior within the protocol.

3.3.1 The PGP trust model

Pretty Good Privacy or *PGP*, is an important milestone in the history of cryptography, because for the first time it makes cryptography available to a wide community. *PGP* was principally created for encrypting or signing e-mail messages and offers a hybrid cryptosystem. In a public cryptosystem it's not necessary to protect public-keys form disclosure. Actually, public-keys ought to be widely accessible by all network participants for encryption. But it is very important to protect public keys from tampering, in order to assure that a public-key really belongs to the person to whom it appears to belong.

Pretty Good Privacy *(PGP)* [27] supports the idea, that all users operate as autonomous certification authorities, which gives them the authorization to sign and verify keys of other entities. The absence of a central trusted third party *(TTP)* was the innovation in this model. The introduction of the decentralized *Web of Trust* allows each entity to sign other keys in order to build a set of virtual interconnections of trust. For example, A knows that B's public-key certificate is authentic and signs it with its private-key. In the following, C wants to communicate with B privately and B forwards its signed certificate to C. C trusts A and finds A among B's certificate signers. Therefore, C can be sure that B's public-key is authentic. However, had C not trusted any of B's certificate signers, including A, C would be skeptical about the authenticity of B's public-key and B would have to find another network participant whom C trusts to sign its public-key certificate. Generally, *PGP* uses the terminology that if A signs B's public-key then A becomes an introducer of B's key. As this process goes on, it establishes the *Web of Trust.* Public-keys certificates are essential to *PGP* and are indispensable to bind the public-key to a network member. Each certificate contains the key owner's user ID, the public-key itself, a unique key ID and the time of creation. Everything may be signed by any number of network participants.

Trust is introduced into the *PGP* Model at two different points, mirrored in the terms: *confidence* and *trustworthiness.* Firstly, *PGP* combines three levels of *confidence* from "undefined" to "marginal" and to "complete" *Trust* for the *trustworthiness* of public-key certificates. This value defines whether a *PGP* public-key certificate is reliable or not in it's binding between the ID and the public-key itself. Secondly, four levels of *trustworthiness* to a public-key are assigned, ranging from "don't know", "untrustworthy" and "marginal" to "full" trust. This value means how much for example C thinks B as the owner of the public-key can be trusted to be the signer or introducer for another public-key certificate. *PGP* requires one "completely" trusted signature or two "marginal" trusted signatures to mark a key as valid.

However, why is *PGP* not suitable for mobile ad-hoc networks even though it sounds obvious that this *Trust Model* might be applied to the idea of decentralized systems even without the existence of a centralized certification authority?

Although the establishment of a central certification authority in the *PGP* model is not necessary, because public-keys are established and signed by network participants themselves, the distribution of public keys is based on continuously accessible public-key directories that reside on centrally managed servers. For this reason, *PGP* is not well applicable for mobile ad-hoc networks where nodes interconnect in an arbitrary way. Additionally, in MANETs nodes form and leave the network dynamically and therefore it is not possible to determine nodes that act as always available public-key certificate servers.

For this reason *PGP* is suitable for wired networks, where this central key server or more central key servers can maintain all keys in a secure database. But the dynamic of wireless links in mobile ad-hoc networks and their spontaneous topology make *PGP* not applicable in MANETs.

Applying an adjusted PGP Trust Model in MANETs

Although PGP public-keys are issued by the participants of the network themselves, the distribution of public-keys is based on uninterrupted and accessible public-key directories that reside on centrally managed servers.

In [14] Jean-Pierre Hubaux, Levente Buttyan and Srdjan Capkun extend the design of *PGP* by establishing a public-key distribution system that better fits to the *self-organized* nature of mobile ad-hoc networks. Similar to *PGP*, public-key certificates are issued, signed and verified by nodes in the MANET themselves based on their individual acquaintances. But, in contrast to *PGP* no continuously accessible public-key directories for the distribution of public-key certificates are necessary. As a substitute, public-key certificates are stored and distributed by the nodes themselves. The main idea is that each node maintains a public-key certificate storage area, called *local certificate repository* containing a subset of public-keys of other entities in the MANET.

The relationships between nodes are represented as a directed graph, called *Trust Graph* containing all nodes of the network. The vertices characterize the nodes or public-keys and the edges represent the public-key certificates issued by other nodes. For instance, there is a directed edge from vertex A to vertex B if node A issued a public-key certificate to node B. The directed path from vertex A to vertex B corresponds to a public-key certificate chain from node A to node B. Thus, the existence of a public-key certificate chain from node A to node B means that vertex B is reachable from vertex A in the directed graph. The *local certificate repository* of every node in the MANET consists of two parts. One part to maintain all public-key certificates issued by the node itself and the second part to store several selected public-key certificates issued by other nodes in the MANET. This means that each node A stores the outgoing edges in conjunction with the corresponding vertices from vertex A as well as an additional set of selected edges in conjunction with the corresponding vertices of the *Trust Graph*. The set of selected edges and vertices of node A, which is also the *local certificate repository*, is called the *Subgraph* that belongs to node A.

In the event that node A wants to verify the public-key of node B, A and B merge their *local certificate repositories* and A tries to discover a suitable public-key certificate chain from node A to node B in the merged public-key certificate storage area. In view of the *graph model*, A and B merge both *Subgraphs* and in the following A tries to find a path from vertex A to vertex B in the merged *Subgraph*. A and B use the same *Subgraph Selection Algorithm*. After node A has verified B's public-key as valid A can start using B's public-key for example to prove his digital signature.

An important element of this model is the *Subgraph Selection Algorithm* because it influences the performance of the system. One characteristic of the *Subgraph Selection Algorithm* is the size of the *Subgraphs* that it selects. Obviously, the performance of *Subgraph Selection Algorithm* and consequently the performance of the system can be increased by selecting larger *Subgraphs*, but then nodes need more memory to store their *Subgraphs*, which may lead to scalability problems. This shows that the small amount of memory storage of a node and the performance of the *Subgraph Selection Algorithm* are opposite requirements in this model.

The *Shortcut Hunter Algorithm* is introduced as *Subgraph Selection Algorithm*. It assumes that there are a dense number of nodes in a small area in order to provide good performance. Shortcuts are found between nodes to keep the *Subgraphs* small and to reduce the storage space on each node. They are stored into the *local certificate repository* based on the number of the *shortcut certificates* connected to the nodes. A *shortcut certificate* is a certificate that, when removed from the graph makes the shortest path between two nodes A and B previously connected by this certificate strictly larger than two. The algorithm selects a *Subgraph* by computing an out-bound and an in-bound path from node A to node B. Both path selection techniques are similar. However the out-bound path algorithm selects in each round an outgoing edge whereas the in-bound path algorithms selects in each round an incoming edge. In conclusion, a public-key certificate chain from node A to node B is found.

So far, this solution assumes that each user is honest and does not issue falsified public-key certificates. In order to compensate for dishonest users an authentication metric is introduced into the model. In this sense, an authentication metric is a function with two nodes A and B and the *Trust Graph* as input. This function returns a numeric value that represents the assurance with which A can obtain the authentic public-key value of B using the information in the *Trust Graph*.

The big advantage of this solution is the *self-organized* distribution of public-key certificates in the MANET without assuming a continuously accessible public-key directory.

However, before being able to verify a public-key, each node must first build its *local certificate repository*, which is a computationally complex operation. Although this initialization phase is performed very rarely, *local certificate repository* becomes outdated if a large number of public-key certificates are revoked. Consequently, the certificate chains might no longer be valid. Therefore the *local certificate repository* has to be rebuilt. For this reason, due to the limited memory and computational power of communicating devices in MANETs, which mainly consist of Personal Digital Assistants *(PDAs)* or mobile phones and the extensive computational and memory requirements of this *self-organized model*, this model is considered as confining for mobile ad-hoc networks.

Furthermore, while analyzing the *Shortcut Hunter Algorithm* for *Subgraph Selection* it stands out, that during verifying a public-key certificate chain from node A to node B, node A must trust the issuer of the public-key certificate for correctly checking that the public-key in the certificate indeed belongs to node B, mostly because of the fact that node A has to select an incoming edge during the in-bound path algorithms. When public-key certificates are issued by mobile nodes of an ad-hoc network, like in MANETs, this method is very vulnerable to malicious nodes that issue false certificates. In order to minimize this problem the an authentication metric is introduced, allowing to determine the degree of authenticity of a public-key by computing the output of a function *f* which uses two nodes A and B and the *Trust Graph* as input parameters. Function *f* could, for example, return the number of disjoint public-key certificate chains from A to B.

Unfortunately, this assumption is vulnerable to *Sybil Attacks* where a malicious node may generate multiple identities for itself to be used at the same time. By launching a *Sybil Attack* the attacker can pretend that different paths are formed by disjoint nodes, although in reality these paths share at least one node which is the attacker's one. Finally, a disproportionate share of the system can become compromised although public-key certificates are utilized.

3.3.2 Decentralized trust model

In 1996 appearing as pioneers Matt Blaze, Joan Feigenbaum and Jack Lacy supported the idea of *"Decentralized Trust Management"* [4] as an important component of security in network services. The Decentralized Trust Management model was the first system taking a comprehensive approach to trust problems independent of any particular application or service. The main achievement was the construction of a system called *PolicyMaker* for defining policies and trust relationships. Handling of *queries* is the fundamental function of the PolicyMaker with the aim to determine whether a specific public-key has the permission to access certain services according to a local policy. Policies are composed in the special *PolicyMaker Language*. A central authority for evaluating credentials is not necessary. Although locally managed, each entity has the competence to achieve own decisions.

The essential point in this model targets the typical problem that, although the binding of the public-key to a network identity was successfully verified, usually the application itself has to subsequently ensure that this network participant is authorized to perform certain actions or is authorized to access security sensitive data. The application for example looks-up the network identity's name in a database and tries to verify that it matches the required service. The Decentralized Trust Model approach wants to establish a generic method that should facilitate the development of security features in a wide range of application, unlike other systems like for example *PGP*. Therefore, this approach extends the common identity-based certificates, which bind a public-key to a unique identity, by means of reliably mapping identities to actions they are trusted to perform. In this context, the specification of policies is merged with the binding of public keys to trusted actions. Consequently, both questions "Who is the holder of the public-key?" and "Can a certain public-key be trusted for a certain purpose?" are clarified with the Decentralized Trust Model. Basically, each network entity that receives a request must have a *policy* that serves as the ultimate source of authority in the local environment.

Currently, the *PolicyMaker* approach binds public-keys to predicates rather than to the identities of the public-key holders. The *PolicyMaker Language* is provided for the purpose of expressing conditions under which a network participant is trusted to sign a certain action. As a result, a network entity has the ability to distinguish between the signatures of different entities depending on the required services. By this means for instance, network entity A may trust certificates signed by network entity B for small transaction but may insist upon certificates from more reliable network entity C for large transactions.

Basically, the *PolicyMaker* service appears to applications like a *database query engine* and functions as a trust management engine. The input is composed of a set of local policy statements (credentials) as well as a string describing the desired trusted action. After evaluating the input, the *PolicyMaker* system finally returns either a yes/no answer or propositions that make the desired action feasible.

All security policies are defined in terms of predicates, called *filters* that are combined with public-keys. The function of the *filters* is to assure if the owner of the corresponding secret-key is accepted or rejected to perform the desired action. A specific action is considered acceptable, if there is a chain from the policy to the key requesting the action, in which all filters are traversed successfully. The design and interpretation of action descriptions, called *action strings*, is not part or even not known to the *PolicyMaker. Action strings* are interpreted only by the calling application and might present various capabilities as signing messages or logging into a computer system. *Action strings* are accepted or rejected by the *filters.*

Signatures can be verified by any public-key cryptosystem, for instance *PGP*. The main reason for it is, that the *PolicyMaker* system does not verify the signatures by itself and that the associated *action strings* are also application specific. Generally, an application calls the *PolicyMaker* after composing the *action string* and determining the identity, from which the desired action originated. Finally, *PolicyMaker* decides whether the *action string* is permitted according the local security policy. The basic function of the *PolicyMaker* system is to process *queries* composed with the *PolicyMaker Language* of the form:

$$key_1, key_2, \ldots, key_n \ \textbf{REQUEST} \ Action\ String$$

A query is a request for information about the trust that can be placed in a certain public-key. The *PolicyMaker* system processes *queries* based on trust information that is included in *assertions*. Assertions assign authority on keys and are of the form:

Source ***ASSERTS*** *AuthorityStruct* ***WHERE*** *Filter*

In this context, each a credential is a type of *assertion,* which binds a *filter* to a sequence of public-keys, called an *authority structure. Source* indicates the origin of the *assertion* and *AuthorityStruct* specifies the public-key(s) to whom the *assertion* applies. Hence, a *Filter* is the predicate that *action strings* must satisfy for the *assertion* to hold. For example, the following *PolicyMaker* credentials indicate that the source *PGP* key "0x01234567abcdefa0a1b2c4d5e6a4f7" asserts that A's *PGP* key is "0xb0034261abc7efa0a1b2c5d4e6a4a3":

```
pgp:"0x01234567abcdefa0a1b2c4d5e6a4f7"
        ASSERTS
                pgp:"0xb0034261abc7efa0a1b2c5d4e6a4a3"
        WHERE
                PREDICATE=regexp:"From A";
```

There are two types of *assertions*: *certificates* and *policies*. The major difference is that *policies* are unconditionally trusted locally and *certificates* are signed messages binding a particular *Authority Structure* to a *filter*. The *Source* field in a *policy assertion* is the keyword "POLICY", rather than the public-key of an entity granting authority.

While this approach provides a basis for expressing and evaluating trust, it does not consider the simultaneous problem of how to continuously control and manage trust over a longer period of time. These problems are discussed by Brent N. Chun and Andy Bavier in [6], where a layered architecture for mitigating the trust management problem in federated systems is proposed. The authors stress that the *PolicyMaker* approach presumes the existence of secure, authenticated channels, for example using preexisting public-key infrastructure, which makes it inapplicable for trust management in MANETs.

3.3.3 Distributed trust model

The Distributed Trust Model in [1] applies a *recommendation protocol* to exchange, revoke and refresh recommendations about other network entities. Therefore each entity needs its own trust database to store different categories of trust values ranging form -1 (complete distrust) to 4 (complete trust). By executing this recommendation protocol, the network entity can determine the trust level of the target node, while requesting for a certain service. The accordant trust level for a single target node is obtained by computing the average value for multiple recommendations. Although this model does not explicitly target ad-hoc networks it could be used to find the selfish, malicious, or faulty entities in order to isolate them so that misbehavior will result in isolation and thus cannot continue.

3.3.4 Distributed public-key trust model

The core of the Distributed Public-Key Trust Model, examined by Lidong Zhou and Zygmund J.Haas [26] is the use of *threshold cryptography* in order to build a highly secure and available key management service. The difficulty of the establishment of a Certification Authority (CA) for key management in MANETs was mentioned in the introductory paragraph. Obviously, the CA, which is responsible for the security of the entire network, is a vulnerable single point of failure that must be continuously accessible by every node. Threshold cryptography implicates sharing of a key by multiple entities called *shareholders* which are involved in authentication and encryption. In [26] the system, as a whole, has a public-/private-key pair and the private-key is distributed over n nodes. Consequently, a central Certification Authority is not necessary. All nodes in the network know the system's public-key and trust any certificate signed using the corresponding private-key. Additionally, each node has a pubic-/private-key pair and has the ability to submit requests to get the public-key of another node or requests to change its own public-key.

The ingenious idea is that $(t+1)$ out of n *shareholders* have the ability to compute the private-key by combining their partial keys but not less then $(t+1)$. In order to obtain the private-key, $(t+1)$ nodes must be compromised. For the service of signing a certificate, each *shareholder* generates a partial signature for the certificate using its private key share and submits the partial signature to one arbitrary *shareholder*, called *combiner*. With $(t+1)$ correct partial signatures the *combiner* is able to compute the signature for the certificate. In the case of one or more incorrect partial signatures generated by compromised nodes, it is not possible to unnoticeably establish a legal signature for the certificate. Fortunately, the *combiner* has the ability to verify the correctness of the signature by using the system's public-key. However, if the verification fails, the *combiner* tries other sets of $(t + 1)$ partial signatures and continues this process until a verifiably correct signature from $(t+1)$ truthful partial parts can be established.

In order to tolerate *mobile* adversaries and to adapt to changes in the network the Distributed Public-Key Trust Model employs a *share refreshing* method. *Mobile adversaries* have the capacity to temporarily compromise one or more *shareholders* and can then move to the next victim. By this technique an adversary may compromise all *shareholders* and gather more than t or even all private-key shares over an extended period of time. Finally, the adversary would be allowed to generate any valid certificate signed by the private-key. *Share refreshing* allows *shareholders* to compute new private-key shares from their old ones in collaboration, but without disclosing the private-key. The new shares are independent from the old and because of this the adversary cannot combine old with new shares in order to recover the private-key.

Although the model offers strong security, like *authentication* of communicating nodes, it has some factors that inhibit its deployment to mobile ad-hoc networks. The pre-establishment of a distributed central authority requires a huge computational complexity and asymmetric cryptographic operations are known to consume precious node battery power. Additionally, the $(t+1)$ parts of the private key may not be reachable to a node requiring authentication and following asymmetric cryptographic services. Finally, the distribution of signed certificates within the MANET is not sufficiently discussed and questionable.

RSA-Based Threshold Cryptography in MANETs

Levent Ertaul and Nitu Chavan visualize in [8] the potentialities and difficulties of *RSA*-based threshold cryptography in MANETs. The examined *RSA* threshold scheme involves key generation, encryption, share generation, share verification, and share combining algorithm. It employs the *Shamir's t-out-of-n* scheme based on *Lagrange's* interpolation. The central idea of this secret sharing scheme is the construction of a *(t – 1)-degree* polynomial over the field $GF(q)$ in order to allow t out of n entities to construct the secret.

$$f(x) = a_0 + a_1 x + \dots + a_{t-1} x^{t-1} \tag{3.1}$$

The coefficient a_0 is the secret and all other coefficients are random elements in the field. The field is known by all entities and each of the n shares is a pair (x_i, y_i) fulfilling the following condition:

$$f(x_i) = y_i \text{ and } x_i \neq 0 \tag{3.2}$$

With t known shares, the polynomial is uniquely determined and the secret a_0 can be computed. The success of the scheme is based on the fact that using t-1 shares, the secret can be any element of the field and is not determinable.

The *RSA-Based Threshold Cryptography* approach makes use of this secret sharing scheme in the following way. After node A has constructed its public-/private-key pair (e,d), the threshold is determined. If node A has n neighbors then the private-key d is partitioned into n partial keys and the neighbors act as *shareholders*. The threshold t is randomly selected under certain conditions:

$$t \geq (n+1)/2,\ t < n, \text{ where } n \geq 2 \tag{3.3}$$

In the subsequent step *Shamir's secret sharing scheme* is applied to calculate key shares and for combining partial messages. Depending on the type of threshold scheme, the secret, and this is always the coefficient a_0 of the polynomial, is different. For threshold encryption, the coefficient a_0 would be e, while for threshold decryption it would be set to d.

We consider a *RSA-Based Threshold Cryptography* based signature scheme between nodes A and B. At first, node A distributes the key shares together with the x_i – values among its n neighbors acting as *shareholders*. x_i – values are selected by A and are public coordinates. The threshold t is not published to the *shareholders* and A notifies only B about t and its public-key e. Consequently, each neighbor has the ability to calculate the partial key $f(x_i)$. Then, A sends the message M securely to all *shareholders* for partial signature generation. *Shareholders* apply $f(x_i)s$ to M and send the partial signature C_is along with the x_i – values to node B. After obtaining at least t partial signature C_is, B sends t selected C_is to A for recovery of C. B encrypts x_i – values using A's public-key e. In the following, A calculates x_i'-values using *Lagrange interpolation* and sends them back to B. Finally, B combines the x_i'- values to the partial signatures in order to get the original C. With $C^e = M$, node B gets the message M for verification.

Due to the exponential computations, the *RSA-Based Threshold Cryptography* scheme requires lots of computational capacity, bandwidth, power and storage. Thus, the authors stress that this approach is unsuitable in resource-constrained MANETs. Another crucial vulnerability of this system is the fact that the neighbors acting as *shareholders* must not authenticate towards node A, from which they get the message M as well as the x_i – values. If the attacker compromises *n-t* or even more *shareholders* he will be able to fake partial signatures in order to disturb the communication between A and B. Although *RSA-Based Threshold Cryptography* does not need a central party to generate shares, it does not consider the vulnerability of wireless links and does not apply to mobility and the dynamically changing network topology in MANETs.

ECC-Based Threshold Cryptography in MANETs

As a result of previous achievements, Levent Ertaul and Nitu Chavan adapt their idea to *ECC*-based threshold cryptography in [9]. Due to the combination of threshold cryptography and *Elliptic Curve Cryptography*, to securely transmit messages in n shares within mobile ad-hoc networks, the performance of *ECC*-based threshold cryptography is more efficient in comparison to *RSA*-based threshold cryptography.

Table 3.1. Key sizes for equivalent security levels in bits (© 2004 IEEE)

Symmetric	ECC	RSA
80	163	1024
128	283	3072
192	409	7680
256	571	15360

Table 3.1 [20] demonstrates, that key sizes can be selected to be much smaller for *ECC* than for *RSA* achieving the same level of security and protection against known attacks.

Although threshold cryptography is a significant approach to build a key management service by distributing the key among a group of entities, the amount of communication for generating the keys, determining the threshold and generating the share could be beyond the scope of available resources in mobile ad-hoc networks, such as computational power, without even considering the problem of finding out a number of routes of disjoint nodes between the sender and receiver in order to choose a number of n shares. All in all, this method is not well suitable for application in mobile ad-hoc networks.

3.3.5 Subjective logic trust model

Jøsang emphasizes in [17] that public-key certificates alone do not assure *authentication* in open networks including mobile ad-hoc networks, for example because of the missing reliable certification authority acting as a Trusted Third Party. This solution introduces an algebra for the characterization of trust relations between entities. A statement such as: *"the key is authentic"* can only be either true or false but nothing in between. However, because of the *imperfect knowledge* about reality it is impossible to know with certainty wheatear such statements are true or false, consequently it is only feasible to have an *opinion* about it. This introduces the notion of *belief* and *disbelief* as well as *uncertainty*. Therefore, *uncertainty* can bridge the gap in the presence of *belief* and *disbelief*. The relationship between these three attributes can be mathematically formulated as follows:

$$b + d + u = 1, \qquad \{b,d,u\} \in [0, 1]^3 \tag{3.4}$$

where b, d and u designate belief, disbelief and uncertainly.

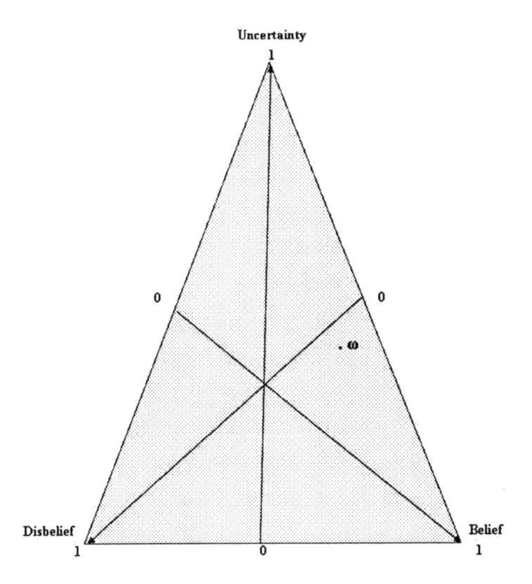

Fig. 3.2. Opinion Triangle

Triples $\omega = \{b, d, u\}$ that satisfy the above condition $b + d + u = 1$ are called *opinions*. Figure 3.2 demonstrates that the condition $b + d + u = 1$ defines a triangle. An opinion ω can be uniquely described as a point $\{b, d, u\}$ in the triangle.

The line between *disbelief* and *belief* corresponds to situations without *uncertainty*. Generally, *uncertainty* is caused by missing evidences in order to either support *belief* or *disbelief*. Obviously, *opinions* are 2-dimensional measures for binary events and binary statements, that either take place or not. *Opinions* are composed by a probability dimension and an *uncertainty* dimension and are according to this determined by *uncertain probabilities*. By mapping the 2-dimensional measures to 1-dimensional probability space a probability expectation value is produced:

$$E(\{b,d,u\}) = b + u/2 \tag{3.5}$$

Opinions of two different entities about the same subject, like for example the binding of a key to an identity, may differ and are not automatically objective. Consequently, the notion of *subjectivity* is introduced in order to express these circumstances. The mathematical technique to characterize *subjectivity* is called *Subjective Logic*. It offers an algebra for determining trust chains by using various logical operators for combing *opinions* that are characterized by *uncertain probabilities*. By enhancing the traditional

Logic, which typically consists of three operators (AND for *conjunction*, OR for *disjunction* and NOT for *negation*), with non-traditional operators such as *recommendation* and *consensus*, the *Subjective Logic* approach is able to deal with *opinions* that are based on other entities' recommendations as well as to produce a single *opinion* about a target statement in the presence of more then one recommendations. As a result, this scheme expands the idea of public-key certificates by introducing trust relations between entities to guarantee *authentication*.

In the following scenario node A receives the public-key of an unknown node B. After ensuring that node B is not included in A's list of *opinions* about the key authenticity, which generally offers an *opinion* about the binding between keys and key owners, and consequently ensuring that B is not included in A's list of *opinions* about the *recommendation* trustworthiness, which explains how much A trusts the key owners to actually recommend keys of other entities, A examines B's public-key certificate. The certificate contains *opinions* about the key authenticity as well as *opinions* about the recommendation trustworthiness assigned by other nodes. Although there might be more than one recommended certification paths to B's key, node A has the capability to determine the authenticity of B's key by computing the *consensus* between the authenticities obtained for each path.

An important assumption of the *Subjective Logic Trust* model is that *opinions*, which are only based on first-hand evidence, should be recommended to other nodes in order to guarantee the independence of *opinions*. Thus, *opinions* based on recommendations from other nodes (second-hand evidence) should never be passed to other nodes. By introducing *uncertainty* in *Trust* it is possible to estimate the consequences of decisions based on trust and recommendations. However, trustworthy authentication of B's public-key requires an unbroken chain of certificates and recommendations. This is a critical condition taking the characteristics of MANETs into account, including the vulnerability to breakage of wireless links and the dynamically changing network topology. Finally, we can conclude that although the *Subjective Logic Trust* approach appears as it needs no Central Trusted Third Party since authenticity of public-keys is based on recommendations, it is not well applicable to mobile ad-hoc networks.

3.4 Recent Trust Models

In this section several state-of-the-art approaches to establish and evaluate *Trust* in mobile ad-hoc networks are presented. The first model [15] introduces the idea to employ an *Ant-based* algorithm in order to compute *Trust Evidences*. The second model [25] focuses on *Trust Evaluation*.

3.4.1 Ant-based trust algorithm

The work of Tao Jiang and John S. Baras [15] presents a scheme for distributing *Trust Certificates* in mobile ad-hoc networks. The core of the model is the *ABED-Ant-Based Evidence Distribution* Algorithm, which is fundamentally based on the *Swarm Intelligence Paradigm* generally used for optimization problems, for example the *Traveling Salesman Problem* (*TSP*) and routing [3]. The major idea of the paradigm is expressed by the term *stigmergy* offering a method for communications in systems by which the individual parts communicate with one another by modifying the environment and without direct interactions. A typical example of *stigmergy* is *pheromone* lying on the paths. Ants, for example, interact with one another by putting *pheromones* along their trails and they follow those trails that have the highest *pheromone* concentration in order to find the optimal path toward their food.

The presented trust model consists of mainly two parts. The first part is the so called *trust computation* model which evaluates the trust level of each entity in the network based on previously retrieved behavioral data or trust evidences. The problem of trust evaluation is not addressed at this point. The second part of a trust model, which is fairly independent of the specific computation of trust, is responsible for trust evidence distribution required for distribution of the calculated trust values to the participating entities. Evidence is presented by trust certificates that are signed by their issuers' private-key. Trust certificates can contain different information depending on the trust model, like for example the public-key or access rights. Trust evidence distribution is a very important subject, because it offers the input for the first part of the trust model, which is the evaluation model.

The main contribution in this work is the reactive *ABED-Ant-Based Evidence Distribution* Algorithm. The procedure starts with several ants that are sent out, when a certain certificate, which serves as a trust evidence of the participating entity, is required. Each node holds its own certificate table, while each entry in this table matches with one certificate. The metric is the probability of choosing a neighbor as the next communicating entity (next hop) instead of the count to destinations.

Two different kinds of forward ants can be mobilized to deliver the required certificate. One of those called the *Unicast ants* that are send out to the neighbor with the highest probability in the certificate table, which means that this neighbor has the required certificate in his certificate table. *Broadcast ants* on the other hand are only sent out when there is no preference to the neighbors, if for example there is no entry in the certificate table for the required certificate in the certificate tables of all neighbors. This can occur in the case if either no path to the certificate has been ascertained or the information is outdated. The density of *pheromone* decides whether the information is valid or outdated. Generally, *pheromone* is utilized in order to route the ants to discover the most favorable path to the required certificate. Furthermore, the decrease of the pheromone density allows the system to update information with the purpose to prevent the outdated information and to look for new paths. The decrease of pheromone is a function of elapsed time, which can be interpreted as a function of mobility. In this manner, a higher mobility means a faster decrease of pheromone. A threshold value τ_0 is defined in order to assure the freshness of the *pheromone*.

Once a forward ant has found the required certificate, a backward ant is generated. This ant retraces the path of the forward ant back to the source and hands the required certificate. By the use of a special *Reinforcement Rule* that is comparable with a learning rule, which is the heart of the ABED, backward ants have the ability to induce certificate table modifications to perform changes. Each node on the path of the backward ant stores the certificate which means that trust certificates get distributed and the certificate table entries of nodes are updated each time the backward ants visit the nodes. A simple *Reinforcement Rule* can be mathematically formulated as follows:

$$P_i(n) = (P_i(n-1) + \Delta p) / (1+ \Delta p)$$
$$P_j(n) = (P_j(n-1) + \Delta p) / (1+ \Delta p) \quad (3.6)$$

$j \in N_k$, where N_k is the neighbor set of node k

i j and i is the neighbor the backward ant came from

$$\Delta p = k / f(c) \quad (3.7)$$

$k > 0$ is a constant and f(c) is a non-decreasing cost-function

Parameter c corresponds to the cost which reveals the information of evidence and could for instance be a measure of hops from the current node to the node where the certificate is located. The model can be enhanced by a *security metric* simply by assigning trust values to paths as costs c. As a result, it will be feasible to draw conclusions: *"the higher the trust value is, the lower is the cost"*. The *Reinforcement Rule* is more complex allowing exploring all information carried by the backward ant and containing the pheromone deposit τ_i.

The main striking question in this approach is, how flexible are ants, particularly backward ants to mobility and especially to link breaks for example if two nodes move far apart?

ABED introduces a special parameter η_j representing the goodness of a link between the current node and its neighbor *j*, which is included in the enhanced *Reinforcement Rule.* In the scenario of link break this parameter is set to a small value and it only assigns a negative reinforcement to the certificate. However, the procedure of finding a secure path from the source to the target node has to be repeated. In a quickly changing MANET environment this solution might lead to long delays. On the other hand, the pheromone, which is used by the ants to mark the crossed path, can be utilized to find a suitable and trustworthy path to the target node quickly.

After simulations of the ABED algorithm and comparisons of the results with those of the P2P Freenet scheme by taking the following three aspects into consideration:

- the number of hops needed to carry the certificate back to the requestor (The cost-function *f(c)* of the *Reinforcement Rule* is the number of hops to the node storing the certificate)
- the delay-time elapsed from sending out the forward ant until receiving the first backward
- the Success Rate measured in percentage of requests for which the requestor successfully receives the certificate

It is observable, that both algorithms converge to the same value, but *I* shows faster convergence at the beginning, which is extremely desired in mobile ad-hoc network setting. Finally, the ABED algorithm outperforms the Freenet-based scheme in the terms of *Success.*

Nevertheless, the *Ant-Based Evidence Distribution Algorithm* assumes that trust certificates are signed by a well known and authenticated signer and that the authentication process takes place prior to the setup of the network. This assumption does not satisfy the nature of mobile ad-hoc networks where nodes may join or leave the network dynamically. Furthermore, allowing new nodes to join the network would implicate the requirement of continuous and secure access to the signer in order to authorize the nodes' public-keys by his signature.

The main weakness of the ABED approach is its vulnerability to Denial-of-Service attacks. Obviously, a malicious and by the attacker compromised node has the capacity to send a huge amount certificate requests for non-existing certificates simultaneously by sending broadcast ants to all its neighbors. Each request will provoke the neighbor nodes to create broadcast ants, because they won't be able to find an entry in their certificate table matching the requested certificate. Consequently, the traffic load increases and may result in a network breakdown.

Furthermore, the attacker may launch a Wormhole attack considering the following scenario, based on the fact that the pheromone deposit which is integrated in the *Reinforcement Rule* and is used to attract ants can only be modified by backward ants. In ABED, backward ants are only generated once a forward ant has found the requested certificate. Then they retrace the path of the forward ant back to the node that has requested the certificate. If the attacker's node behaves unnoticeably and generates unicast and broadcast ants in accordance with the algorithm, forward ants will find the path to the requested certificate and generate a backward ant passing the attacker's node. In the moment where the backward ant reaches the attacker's node and wants to modify its certification table, the attacker might discard the backward ant and may obtain the certificate out of the backward ant's packet. As a result, the requesting node won't be able to receive the certificate as trust evidence.

Finally, the ant-based evidence distribution algorithm offers an innovative approach to obtain a distribution of previously, by the trust model defined, trust values within a network, like a mobile ad-hoc network but on the other hand the algorithm has to deal with a high vulnerability to multiple attacks.

3.4.2 Using cooperative games and distributed trust computation in MANETs

This model [16] demonstrates that dynamic cooperative games provide a natural framework for the analysis of multiple problems in MANETs while concentrating on distributed trust computation in addition to trust distribution, explained 3.4.1. Assuming that trust computation is distributed and restricted to only local interaction, a mobile ad-hoc network is modeled as an undirected graph (V, E) where the edges represent connections to exchange trust information. In this context, it is not necessary that two end-nodes of an edge are neighbors in geometrical distance although they have a trust relationship. The distributed trust computation model is based on elementary voting methods and only nodes in node's neighborhood have the right to vote. By this technique, it is possible to mark a node as trustworthy or not. A secure path in this concept is a path consisting only of trusted nodes.

Unfortunately, this approach is vulnerable to Sybil attacks, where the attacker can represent multiple identities and has then the capacity to generate fake recommendations about the trustworthiness of a certain node in order to attract more traffic to this node.

3.4.3 Using semirings to evaluate trust in MANETs

In [25] a concept on how to establish an indirect trust relationship without previous direct interactions within an ad-hoc network is introduced. By the use of the theory of semirings, the presented approach is also robust in the presence of attackers. The significant idea is to model the trust inference problem as a generalized shortest path problem on a weighted graph G(V,E), also referred to as the *trust graph.* A weighted edge corresponds to the opinion, consisting of two values the *trust* value and the *confidence* value that an entity has about another entity in the graph (network). In this model, a node has the ability to rely on others' past experiences and not just on his own, which might be insufficient, to ascertain if the target node is trustworthy. The problem of finding a trusted path of nodes is also solved in this model. This scheme does not need any centralized infrastructure and can be seen as an extension of the traditional PGP model explained in 3.3.1. The main difference to PGP is that PGP uses only directly assigned trust values, whereas the use of semirings allows entities to compute an opinion about other network entities even without the need of personal or direct interactions with every other user in the network. The model has a strong theoretical framework but is vulnerable to Sybil attacks.

3.5 Conclusions

Security-sensitive data and applications transmitted within mobile ad-hoc networks require a high degree of security. *Trust* as a concept of security services has the ability to achieve the required level of security with respect to mobility and constraints in resources of the participating devices. This chapter presented several trust models, such as PGP as well as new models taking the dynamic and mobile nature of mobile ad-hoc networks into consideration. Altogether, *Trust* as a security principle of as foundations for succeeding security principles, like for example authentication, evolves to become more and more important in mobile ad-hoc network settings. Primarily, the use of trust recommendations and second-hand information, based on trusted relationships, might significantly speed up the discovery of malicious behavior and may consequently facilitate the isolation of malicious nodes in mobile ad-hoc networks. Particularly, the Ant-based Adaptive Trust Evidence Distribution Model provides the required adaptivity to network changes and tolerance of faults in networks and offers a dynamic method to obtain trust evidence in mobile ad-hoc network settings.

Finally, it is noticeable that every *Trust* evaluation, *Trust* computation, and *Trust* distribution model applied in mobile ad-hoc network settings has to struggle with at least one of the following two problems. Either the model requires unrealistic assumptions like for example the requirement of a continuously accessible and centrally managed public-key database or an algorithms with a high computational complexity, or if no unrealistic assumptions exist, the model has a high vulnerability to multiple attacks. As a result, the question: *"Which Trust model is the best in mobile ad-hoc network setting?"* can only be answered with reference to the application or application area the mobile ad-hoc network is established in. Basically, two major application areas for MANETs can be differentiated: public mobile ad-hoc networks and mobile ad-hoc networks in military or emergency scenarios. Depending on the application area, different threats might vulnerate the mobile ad-hoc network. Choosing for example the public network application scenario, the most important function of the *Trust* model will be to facilitate cooperative behavior between all entities in the MANET. Hence, the biggest threats to the *Trust* model and the MANET pose selfish behavior and Denial-of-Service attacks. For that reason, recommendation based *Trust* models, like for example the *Distributed Trust Model* (3.3.3) can introduce a reputation system into the mobile ad-hoc network, allowing the isolation of selfish and unfair network entities. In emergency and military application scenarios, the main threat to the mobile ad-hoc network pose attacks targeting the privacy and confidentiality of communications. However, in these crisis situations the communicating entities are known to each other, which means that there exists the possibility to exchange secure-keys or even to distribute public-keys previous to communication. Therefore, encryption based *Trust* models, such as the *Distributed Public-Key Trust Model* (3.3.4), can be deployed in order to preserve confidential communications. Obviously, mobile ad-hoc networks in crises situations require a much stronger protection than in the public network setting since human life needs to be protected. Consequently, the *Trust* condition has to be followed rigorously extended by additional mobile ad-hoc network security technologies.

In conclusion, the establishment of *Trust* in mobile ad-hoc network settings is an extremely powerful method offering an additional component of security principles or functioning as an essential foundation for succeeding security terms, such as authentication. Nevertheless, introducing *Trust* for mobile ad-hoc network protection has to be implemented with care and always in accordance with the characteristics and vulnerabilities of the application area.

References

1. Abduhl-Rahman A, Hailes S (1997) A distributed trust model. In: Workshop on New security paradigms. United Kingdom, pp 48–60
2. Awerbuch MB, Curtmola R, Holmer D, Nita-Rotau Cristina, Rubens Hubert (2004) Mitigating Byzantine Attacks in Ad Hoc Wireless Networks. In: Technical Report Version 1
3. Awerbuch MB, Holmer D, Rubens Hubert (2004) Swarm Intelligence Routing Resilient to Byzantine Adversaries. In: International Zurich Seminar on Communications. pp 160–163
4. Blaze M, Feigenbaum J, Lacy J (1996) Decentralized Trust Management. In: IEEE Symposium for Security and Privacy. USA: 96-17: 164
5. Buchegger S, Le Boudec JY (2005) Self-Policing Mobile Ad-Hoc Networks by Reputation. In: IEEE Communication Magazine. USA: 43-7: 101–107
6. Chun B, Bavier A (2005) Decentralized Trust Management and Accountability in Federated Systems. In: 37th Annual Hawaii International Conference on System Sciences. USA
7. Douceur JR (2002) The Sybil Attack. In: IPTP02 Workshop. Cambridge, MA, USA
8. Ertaul L, Chavan N (2005) Security of Ad Hoc Networks and Threshold Cryptography. In: International Conference on Wireless Networks, Communications, and Mobile Computing (WirelessCom). USA: pp 69–74
9. Ertaul L, Weimin L (2005) ECC Based Threshold Cryptography for Secure Data Forwarding and Security Key Exchange in MANET (I). In: NETWORKING 2005. LNCS Berlin: 3462/2005: pp 102–113
10. Eschenauer L, Gligor VD, Baras JS (2002) On trust establishment in mobile ad-hoc networks. In: LNCS Berlin: 2845/2003: pp 47–66
11. Hu YC, Perrig A, Johnson DB (2002) Ariadne: A secure on-demand routing protocol for ad hoc networks. In: Wireless Networks. Springer Netherlands: vol 11, 2005: pp 21–38
12. Hu YC, Perrig A, Johnson DB (2003) Packet Leashes: A Defense against Wormhole Attacks in Wireless Ad Hoc Network. In: IEEE INFOCOM 2003. USA: vol 3, pp 1976–1986
13. Hu YC, Perrig A, Johnson DB (2003) Rushing Attacks and Defense in Wireless Ad Hoc Networks. In: ACM Workshop on Wireless Security. USA: pp 30–40
14. Hubaux JP, Buttyan L, Capkun S (2001) The Quest for Security in Mobile Ad Hoc Networks. In: 2nd ACM international symposium on Mobile ad hoc networking & computing. USA: pp 146–155
15. Jiang T, Baras J (2004) Ant-based Adaptive Trust Evidence Distribution in MANET. In: 2nd International Workshop on Mobile Distributed Computing. USA: pp 588–593
16. Jiang T, Baras J (2004) Cooperative Games, Phase Transition on Graphs and Distributed Trust MANET. In: 43rd IEEE Conference on Decision and Control. Bahamas

17. Jøsang A (1999) An Algebra for Assessing Trust in certification Chains. In: Network and Distributed Systems Security Symposium (NDSS 99). The Internet Society. USA
18. Jøsang A, Keser C, Dimitrakos T (2005) Can We Manage Trust? In: LNCS Berlin: 3477/2005 pp 93–107
19. Lamsal P (2001) Understanding Trust and Security. In: Department of Computer Science. University of Helsinki. Finland
20. Lauter K (2004) The Advantages of Elliptic Curve Cryptography for Wireless Security. In: IEEE Wireless Communications. USA: vol 11 pp 62–67
21. Pirzada AA, McDonald C (2004) Establishing trust in pure ad-hoc networks. In: 27th Australasian conference on Computer science. New Zealand: vol 26 pp 47–54
22. Ramaswamy S, Fz H, Sreekantaradhya M, Dixon J, Nygard K (2003) Prevention of Cooperative Black Hole Attack in Wireless Ad Hoc Network. In: International Conference on Wireless Networks. USA: pp 570–575
23. Sanzgiri K, Dahill B, Levine B, Belding-Royer E (2002) A secure routing protocol for ad hoc networks. In: International Conference on Network Protocols. France
24. Spiewak D, Engel T, Fusenig V (2007) Unmasking Threats in Mobile Wireless Network Settings. In: WSEAS Transactions on Communications vol 6 pp 104–110
25. Theodorakopoulos G, Baras J (2004) Trust Evaluation in Ad-Hoc Networks. In: ACM workshop on Wireless security. USA: pp1–10
26. Zhou L, Haas ZJ (1999) Securing Ad Hoc Networks. In: IEEE Network special issue on network security. USA: 13(6): 2430
27. Zimmermann PR (1995) The Official PGP User's Guide. MIT Press USA

4 A Framework for Computing Trust in Mobile Ad-Hoc Networks

Tirthankar Ghosh[1], Niki Pissinou[2], Kia Makki[2], and Ahmad Farhat[2]

[1] Computer Networking and Applications, College of Science and Engineering, St. Cloud State University, St. Cloud Minnesota USA
[2] Telecommunications and Information Technology Institute, College of Engineering and Computing, Florida International University, Miami, Florida, USA

4.1 Introduction

Modeling and computing trusts in ad-hoc networks is a challenging problem. It is very difficult to form a true and honest opinion about the trustworthiness of the nodes in such applications where the network is formed with near-strangers relying on one another for normal network operation without any prior knowledge of trustworthiness. These near-strangers can be engaged in malicious activities in different ways. This intricacy in trust computation, together with frequent topology changes among nodes, quite often causes the whole network to get compromised or disrupted. Different malicious activities of the nodes can very well be misinterpreted as the regular erratic behavior of the wireless networks in general and ad-hoc networks in particular, thus making trust computation even more difficult. In this paper we have proposed a framework for modeling and computing trusts that take into account different malicious behavior of the nodes. Our proposed model tries to explore the behavioral pattern of the attacker in different ways and quantifies those behaviors to form a computing framework.

In some of the earlier works on trust computation, incentive mechanisms have been proposed to prevent selfish behavior among the nodes. These mechanisms can be either reputation-based incentive mechanisms (Buchegger and Boudec 2002; Michiardi and Molva 2002), or price-based incentive mechanism (Buttyán and Hubaux 2003). In both the mechanisms, nodes are given incentives to suppress their malicious intention in favor of the network. But nodes with malicious intentions always try to find ways to bypass these incentive mechanisms. In our work, instead of forcing the nodes to act in an unselfish way, we propose to develop a trust model by collaborative effort and use this model in the trusted routing solution proposed by us in our earlier work [Pissinou et al. 2004; Ghosh et al. 2004].

4.2 Related Work

Establishing security associations based on distributed trust among nodes in an ad-hoc network is an important consideration while designing a secure routing solution. Although some work has been done lately to design trusted routing solution in ad-hoc networks, not much work has been done to develop a trust model to build-up, distribute and manage trust levels among the ad-hoc nodes. Most of the proposed schemes talk about the general requirement of trust establishment (Verma et al. 2001; Eschenauer 2002; Kagal et al. 2001; Lamsal 2002; Buchegger and Boudec 2002). Some work has been done to propose models for building up trust (Jiang and Baras 2004; Theodorakopoulos and Baras 2004), but they do not specify the detailed incorporation of different malicious behavior in those models. In (Theodorakopoulos and Baras 2004) the authors proposed a trust establishment model based on the theory of semirings. A trust distribution model has been proposed in (Jiang and Baras 2004) using distributed certificates based on ant systems. However, none of the models proposed so far have tried to analyze the behavioral pattern of the attacker and quantify those behaviors in the computational framework.

Modeling and computing trust for a distributed environment has been actively researched for quite sometime (Beth 1994; Abdul-Rahman and Hailes 1997). Most of these distributed trust models combine direct and recommended trusts to come up with some sort of trust computations.

Watchdog mechanism (Marti et al. 2000), based on promiscuous mode operation of the ad-hoc nodes, has been the fundamental assumption in any trust computational model. In (Yan et al. 2003) the authors have proposed a trust evaluation-based secure routing solution. The trust evaluation is done based on several parameters stored in a trust matrix at each ad-hoc node. However, the mechanism for collecting the required parameters was not discussed by the authors. Also, some of the parameters suggested by the authors are not realistic in a highly sensitive application. In (Pirzada and McDonald 2004) the authors have proposed a model for trust computation based on parameter collection by the nodes in promiscuous mode. However, the trust computation is based only on the success and failure of transmission of different packets and does not take into account different forms of malicious behavior.

In (Ngai and Lyu 2004) the authors have proposed an authentication scheme based on Public Key infrastructure and distributed trust relationship. The trust relationship is established by direct as well as recommended trusts. Composite trust is computed by combining both direct and recommended trust relationships.

Some work has also been done to establish trust based on distribution of certificates. In (Davis 2004) the authors have proposed such a trust management scheme. Trust revocation is done by carrying out a weighted analysis of the accusations received from different nodes. However, the proposed scheme lacks any specific framework for computing the indices.

Another model has been proposed based on subjective logic (Li et al. 2004). The concept of subjective logic was first proposed in (Josang 2001, 1998, 1997). Subjective logic is "*a logic which operates on subjective beliefs about the world, and uses the term opinion to denote the representation of a subjective belief*" (Josang 2001). An opinion towards another entity x is represented by three states: *belief* $[b(x)]$, *disbelief* $[d(x)]$ and *uncertainty* $[u(x)]$, with the following equality:

$$b(x) + d(x) + u(x) = 1$$

The concept of subjective logic has been extended to propose a trusted routing solution in (Li et al. 2004). Each node maintains its trust relationships with neighbors, which are updated depending on positive or negative impression based upon successful or failed communication with neighboring nodes. The opinion of a node about another node is represented in a three-dimensional metric representing trust, distrust and uncertain opinions. However, this scheme fails to save the network from an internal attack, where a malicious node either refuses to forward the packets and duly authenticates itself to the source, or it cooperates with the source node and acts as a black hole.

Some mechanisms have been proposed to give incentives to the nodes for acting unselfishly. In (He et al. 2004) authors have proposed a secure reputation-based incentive scheme (SORI) that prevents the nodes from behaving in a selfish way. The scheme, however, does not prevent a malicious node from exhibiting other malicious behavior.

4.3 Proposed Model

4.3.1 Understanding different malicious behavior

Our motivation for developing the trust model is to have a framework to form a true opinion about the trustworthiness of the nodes by analyzing various malicious behavior. To do this we need to understand clearly the ways a node can engage itself in different malicious acts. Below we highlight the different malicious behavior.

- A node engaging in selfish behavior by not forwarding packets meant for other nodes.
- A node falsely accusing another node for not forwarding its packets, thus isolating the node from normal network operation.
- A node placing itself in active route and then coming out to break the route (route flapping), thus forcing more route request packets to be injected into the network. By repeating this malicious act, a large number of routing overhead is forcefully generated wasting valuable bandwidth and disrupting normal network operation.

4.3.2 The model

Our model has been developed with a view to form an honest opinion about the trustworthiness of the nodes with collaborative effort from their neighbors. The underlying assumptions in developing the model are the existence of shared bi-directional wireless channels, promiscuous operation of the ad-hoc nodes, and the existence of an on-demand routing protocol on top of which our proposed model can be built. In the following section we analyze different malicious behavior and quantify them to gradually develop the model.

4.3.2.1 Trust Model Against Selfish Behavior

The development of the model to punish a node for selfish behavior is based on the Secure and Objective Reputation-based Incentive (SORI) scheme proposed in (He et al. 2004) with several modifications. We will elaborate more on these modifications as we describe the trust model. The parameters are described below:

(i) NNL_N= Neighbor Node List (each node maintains a list of its neighbors, either by receiving *Hello* messages, or by learning from overhearing).

(ii) $RF_N(X)$(Request for Forwarding) = total number of packets node N has forwarded to node X for further forwarding.

(iii) $HF_N(X)$(Has Forwarded) = total number of packets that have been forwarded by X and noticed by N.

We are not discussing the details of updating these parameters, which can be found in (He et al. 2004). With the above parameters, node N can create a *local evaluation record* (denoted by $LER_N(X)$) about X. The record $LER_N(X)$ consists of two parameters shown below:

$LER_N(X)$= Local Evaluation Record of node N of node X. It reflects the evaluation of the behavior of node X by another node N.

where,

$G_N(X)$= Forwarding ratio of node N on node X.

$C_N(X)$= Confidence level of N on X.

The confidence level $C_N(X)$ is computed as below:

$$C_N(X) = \sum_t HF_N(X) / \sum_t RF_N(X) \tag{4.1}$$

Node N computes its confidence level on X after sending packets to X over a time period t.

We propose a similar propagation model proposed in SORI. Each node updates its local evaluation record (LER) and sends it to its neighbors. When a node N receives the $LER_i(X)$ from node i, it computes the overall evaluation record of X (denoted by $OER_N(X)$), as given below:

$$OER_N(X) = \frac{\sum_{i \in NNL, i \neq X} C_N(i) * C_i(X) * G_i(X)}{\sum_{i \in NNL, i \neq X} C_N(i) * C_i(X)} \tag{4.2}$$

where, $C_N(i)$ = confidence level of node N on node i from which it receives $LER_i(X)$

$C_i(X)$ = confidence level of node i on node X

$G_i(X)$ = forwarding ratio of node i on X

4.3.2.2 Trust Model Against Malicious Accuser

The calculation of confidence level in equation 4.1 is based only on the nodes' decision to forward packets, and does not take into account the malicious accusation of a node about another node. We foresee a threat where a node falsely accuses another node of not forwarding its packets, eventually to isolate the later as an untrustworthy node. This malicious act should also be reflected in the trust computation, where every node should be given a chance to defend itself. Equation 4.3 below shows the calculation of confidence level taking into account both selfish behavior and false accusation.

$$C_N(X) = (\sum_t HF_N(X) / \sum_t RF_N(X)) * \alpha_X(N) \quad (4.3)$$

where, $\alpha_x(N)$ = *accusation index* of N by X

$$= \begin{cases} 0; & \text{if X falsely accuses N} \\ 1; & \text{otherwise} \end{cases}$$

Node N keeps a track of the packets it received from X and packets it forwarded. If N finds out that X is falsely accusing it for non-cooperation, it recomputes its confidence level on X by taking into account the accusation index. It then broadcasts the new $LER_N(X)$ with new $C_N(X)$, thus resulting in computation of a new $OER_N(X)$, which is low enough to punish X. Thus, any sort of malicious behavior of X by falsely accusing other nodes gets punished eventually.

4.3.2.3 Conflict Resolution

It may so happen that two nodes come up with conflicting views of each other. This can be a common problem in ad-hoc networks as the nodes are forced to communicate with near-strangers without any prior information about their trustworthiness. To resolve such conflicting views and compute an honest opinion of a node's trustworthiness, we need to consider three scenarios.

- Scenario I: Two nodes have mutual high trust of each other: this scenario does not lead to a conflicting opinion, and can be treated as the normal and expected behavior. However, if two nodes collude with each other and come up with high mutual trust, while not cooperating with other nodes, this can lead to an untrustworthy situation that severely affects network performance and security. However, the discussion and analysis of colluding threats are beyond the scope of this paper.
- Scenario II: Two nodes have mutual low trust of each other: if both the nodes are to be believed, then they are to be isolated as malicious and non-cooperating nodes. The network will be safer, but the decision will affect network performance. This can be viewed as a conservative approach.
- Scenario III: Two nodes have conflicting opinions about each other: this scenario can lead to two different cases. First, if both the nodes are right in assessing each other, then one of them is not cooperating;

and second, if one of them is falsely accusing the other, that will lead to the malicious accusation scenario discussed earlier. However, both these two cases will ultimately lead to scenario II, as the node getting accused (falsely or rightly) will eventually accuse its accuser, and both will have low trust of each other.

Resolving this type of conflict is non-trivial. When a node receives mutual low confidence of other nodes, it has two clear choices: either to believe both, or to believe one of them. If both nodes are to be believed, then they are barred from taking part in the route selection process, essentially isolating them from participating in the normal network operation. This approach is viewed as extremely conservative, and, although secure, will degrade the network performance as more nodes start getting isolated. However, we have a different approach to solve the conflicting situation. In our approach when a node receives mutual low confidence of other nodes, it will put both of them in quarantine, and will monitor their behavior without changing their confidence levels. However, if the quarantined nodes persist with mutual low trust, and their assessment by other nodes start getting low, they are eventually isolated. On the other hand, if the nodes change their mutual opinions, they are removed from the quarantine. This acts as an incentive to a malicious accuser for not accusing other nodes falsely, because that will eventually isolate the accuser too, which will defy its purpose of accusing other nodes. The amount of time the nodes will be in quarantine is a critical design parameter that will affect the overall network performance, the discussion of which is beyond the scope of this paper.

4.3.2.4 Trust Model Against Malicious Topology Change

A node may engage in route flapping, where it forces the network topology to change frequently by putting itself in active route and then withdrawing and putting itself back. This will generate a large number of route request packets, essentially slowing down the network operation. If such a behavior is detected, the confidence level must be changed in order to punish the malicious node. However, detection of such a behavior is not easy, as any such topology change can be viewed as a normal characteristic of an ad-hoc network. We have tried to capture such a malicious act by modeling the action and reflecting it in the computation of trust.

To develop the model, we require each node to maintain a table called a *neighbor remove table*, where it keeps track of any node moving out of the path. The table is populated by successive *Hello* misses in AODV (Perkins and Royer 1999), or from the *unreachable node address* field in the RERR packet in DSR (Johnson and Maltz 1999). A snapshot of the table is shown below:

Table 4.1. Snapshot of Neighbor Remove Table

Node Address	Time of Leaving	Time Difference
X	T1	t0 = 0
X	T2	t1 = T2 – T1
X	T3	t2 = T3 – T2
X	T4	t3 = T4 – T3
		Mean = μ_t

Each node periodically scans the table to find whether any particular node is leaving at frequent intervals. It computes the mean, μ_t of the time difference of any particular node leaving the network. If μ_t is found lower than a threshold value (denoted by $t_{threshold}$), then the node is identified as malicious and the confidence level is computed as follows:

$$C_N(X) = (\sum_t HF_N(X) / \sum_t RF_N(X)) * m(X) \tag{4.4}$$

where, *m(X)* = *malicious index* of node X

$$= \begin{cases} 0; & \text{if } \mu_t <= t_{threshold} \\ 1; & \text{otherwise} \end{cases}$$

The choice of the threshold value can be selected based on the application for which the ad-hoc network is deployed. A network that demands frequent topology change can have a higher threshold to accommodate the normal network behavior. The choice is not discussed in this paper and is left for future consideration.

Finally, to combine all the malicious behavior discussed earlier and to reflect those behavior in trust computation, the confidence level of node N on X is computed as shown below:

$$C_N(X) = (\sum_t HF_N(X) / \sum_t RF_N(X)) * \alpha_X(N) * m(X) \tag{4.5}$$

The final overall evaluation record (OER), when computed based on the local LERs, will reflect the different malicious behavior of a node as computed in the confidence level, and finally any malicious act gets detected and punished.

4.4 Simulation

Although simulation is not quite foolproof, and only implementation in a real environment can assure us of the effectiveness of any design, it is extremely difficult to test proposed protocols and designs on a large-scale real-life ad-hoc network testbed due to lack of availability. Like all other evaluation methods, we have also reverted to simulation for evaluating our trust model. We have used Glomosim (Zeng et al. 1998) for our simulation. Glomosim is a scalable simulation software used for mobile ad-hoc networks. We have carried out the simulation with two different scenarios. We defined a region of 2 Km by 2 Km and placed the nodes randomly within that region. In the first scenario, the nodes moved with uniform speed chosen between 0 to 10 meters/sec with 30 seconds pause between each successive movement. We increased the number of nodes and studied the network performance. In the second scenario, we have increased the node speed, keeping the similar infrastructure, to carry out our analysis. The parameters for both the scenarios are shown in the table below.

Table 4.2. Parameters used for simulation

	Independent variable	**Set of parameters compared**			
1	Number of nodes	Routing overhead	Number of route errors	Throughput	Average end-to-end delay
	Independent variable	**Set of parameters compared**			
2	Node speed	Routing overhead	Number of route errors	Throughput	Average end-to-end delay

We have incorporated trust computation directly into the routing protocol to avoid any unnecessary layering interoperability. We have extended the Ad-hoc On-Demand Distance Vector (AODV) routing protocol (Perkins and Royer 1999) to incorporate the trust computation and exchange. The modified protocol has been benchmarked with base AODV to study its scalability and efficiency. To avoid any unwanted overhead we have ensured the trust information exchange to be piggybacked with the route request packet header. From Fig 4.1 we can see that our protocol scales as good as the base AODV with increasing number of nodes. Even though we have incorporated extensive trust computation at each node both by its own spying mechanism as well as by exchanging information from its neighbors, we can see that our protocol does not add any significant overhead.

Similar results can be seen from Fig 4.2 where we have benchmarked our modified protocol with AODV in terms of route errors sent. Number of route errors are dependent on several factors like localized clustering of the nodes, MAC layer load and also routing and transport layer load. The parameters show random variation as quite expected from the ad-hoc nature of the whole network. In both the cases we can see that the modified protocol scales as good as AODV even with large network size.

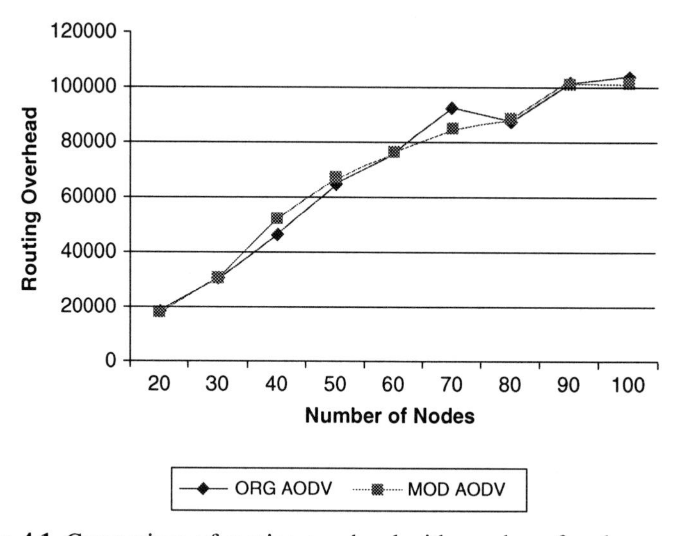

Fig. 4.1. Comparison of routing overhead with number of nodes

Figs 4.3 and 4.4 compare the average end-to-end delay (in seconds) and throughput (in bits per second) respectively for the base AODV and our modified protocol. It can be concluded that our modified protocol scales as good as the base AODV with respect to these parameters as well. These parameters also depend upon the localized clustering of the ad-hoc nodes and overall network load including MAC layer, network layer and transport layer loads. Hence these parameters also show random variation for the two protocols.

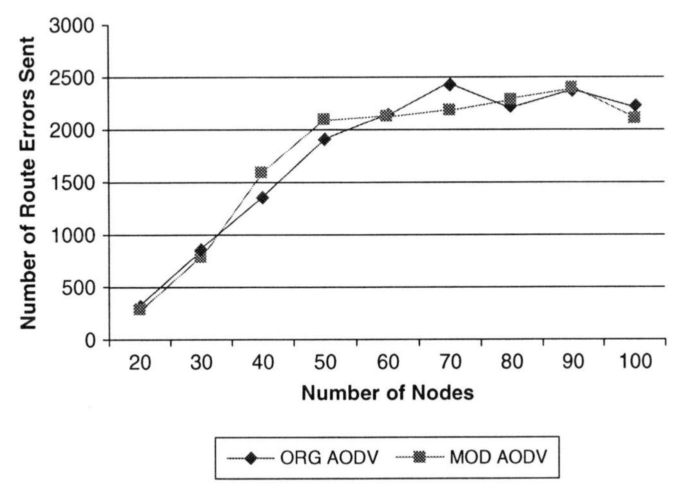

Fig. 4.2. Comparison of route errors with number of nodes

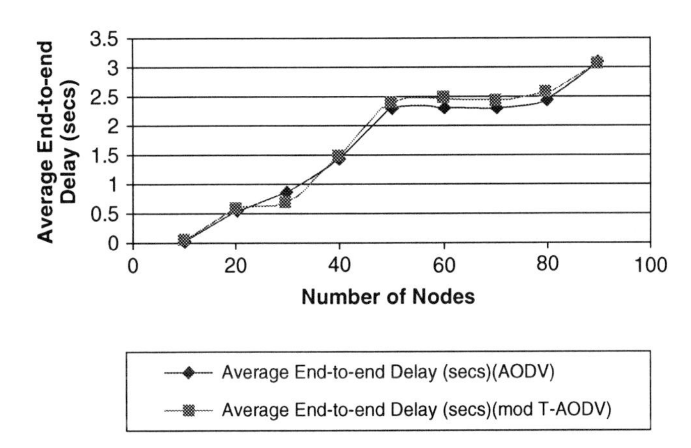

Fig. 4.3. Comparison of average end-to-end delay with number of nodes

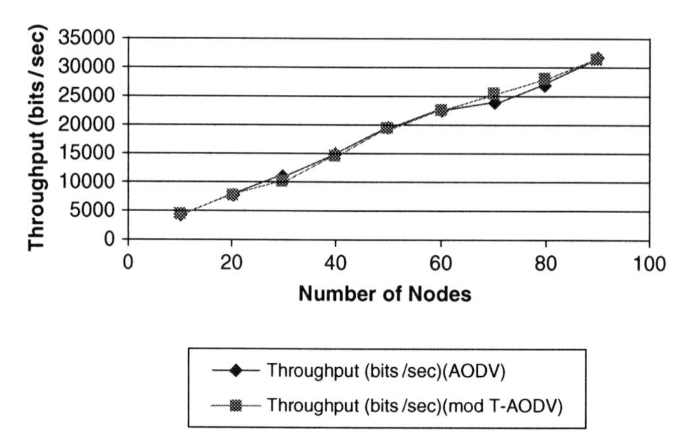

Fig. 4.4. Comparison of throughput with number of nodes

Our next set of simulation is to evaluate the modified protocol with increasing node speed. This parameter has been selected to see the protocol scalability and efficiency with frequent changes in network topology. We can see from Fig. 4.5 that our modified protocol does not add any overhead, even with higher node movement. Fig. 4.6 concludes in a similar way that the protocol scales very well in terms of route errors sent.

As we have piggybacked the confidence information into the route request messages to control routing overhead, we can conclude that mobility will help in updating trust and confidence information in our modified protocol. As the topology of the network changes more frequently necessitating more and more route request packets to be generated, more recent information about the trusts are circulated in the network. Thus, we can conclude that our modified protocol is not only efficient and scalable with network size and node speed, it also gives a better picture of trust and confidence with higher node speed.

Figure 4.7 compares the average end-to-end delay (in seconds) for the base AODV and the modified protocol. We can see that the modified protocol scales as good as the original AODV with increasing node speed with respect to the delay.

As we can see from the simulation, the parameters for the modified protocol vary randomly with comparison to the base AODV with sometimes lower and sometimes higher values. This is attributed mainly to the ad-hoc nature of the network with random waypoint mobility model. The

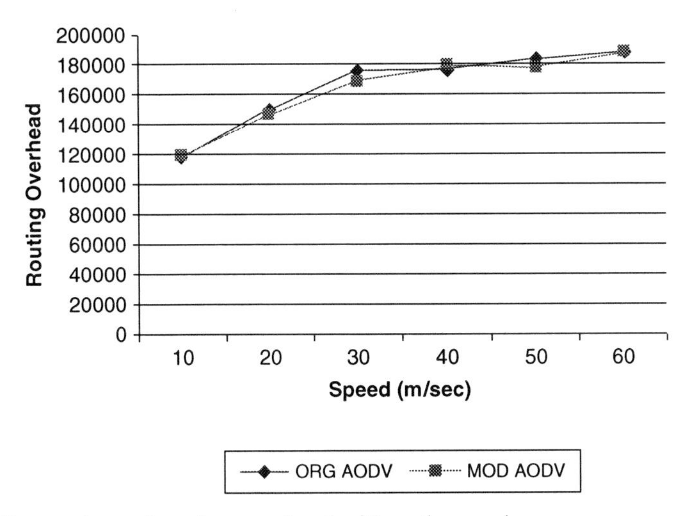

Fig. 4.5. Comparison of routing overhead with node speed

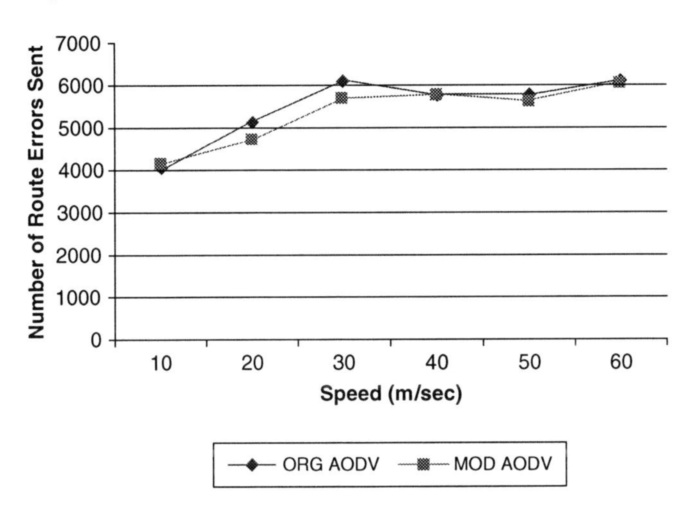

Fig. 4.6. Comparison of route errors with node speed

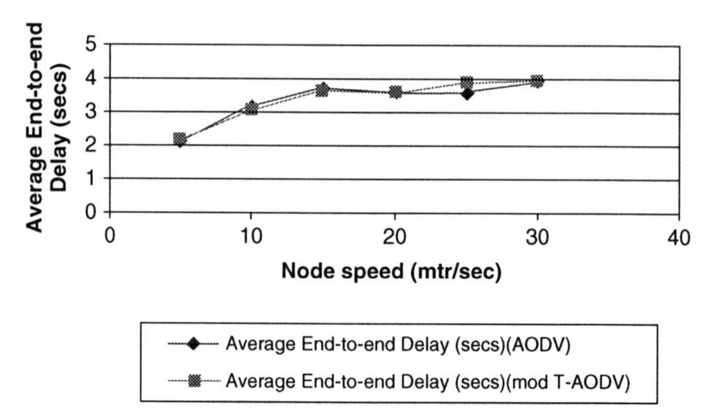

Fig. 4.7. Comparison of average end-to-end delay with node speed

parameters are dependent upon factors like localized node clustering, MAC layer load and also transport and network layer load, as we have discussed previously. These factors change with every simulation run with random waypoint mobility, which attributes to the somewhat random variation between the two protocols.

4.5 Conclusion

We have developed a model for trust computation in mobile ad-hoc networks based on different malicious behavior of the nodes. Our model is unique in the sense that it tries to explore different malicious behavior of the attacker and quantifies those behaviors to form a computing framework, where any malicious act eventually gets detected. This model for computing and updating trusts is to be integrated with the trusted routing protocol proposed by us (Ghosh et al. 2005, 2004) to come up with a secure and robust routing solution that can efficiently withstand attacks from malicious nodes acting either independently or in collusion.

Although our proposed model forms a foundation for trust computation based on different malicious behavior in an ad-hoc network, we feel that there is much to be done in this area. More malicious behaviors need to be identified and quantified into the model. Furthermore, as we have discussed before, our simulation provides only a basic foundation for evaluating the proposed trust model and the trusted routing protocol; and we need to carry out a detailed experimental analysis on a large-scale real-time ad-hoc testbed. Currently we are in the process of creating such a testbed in our lab, which will eventually be extended in a wide-area setting, and the proposed model will be tested in that testbed. This will give us a more realistic evaluation of our model.

References

1. Abdul-Rahman A, Hailes S (1997) A Distributed Trust Model. ACM New Security Paradigm Workshop, 1997.
2. Aberer K, Despotovic Z (2001) Managing Trust in a Peer-2-Peer Information Systems. In: Proceedings of Conference on Information and Knowledge Management (CIKM). Atlanta, Georgia, USA.
3. Beth T, Borcherding M, Klein B (1994) Valuation of Trust in Open Networks In: Proceedings of the European Symposium on Research in Computer Security (ESORICS). Brighton, UK, pp 3–18.
4. Blaze M, Feigenbaum J, Lacy J (1996) Decentralized Trust Management. In: Proceedings of the IEEE Conference on Security and Privacy. Oakland, CA, USA.
5. Buchegger S, Boudec JL (2002) Nodes Bearing Grudges: Towards Routing Security, Fairness, and Robustness in Mobile Ad-hoc Networks. In: Proceedings of the Tenth Euromicro Workshop on Parallel, Distributed, Network-based Processing. Canary Islands, Spain, pp 403–410.
6. Buchegger S, Boudec JL (2002) Performance Analysis of the CONFIDANT Protocol (Cooperation Of Nodes: Fairness In Dynamic Ad-hoc Networks). MOBIHOC, Switzerland.
7. Buttyán L, Hubaux JP (2003) Stimulating Cooperation in Self-Organizing Mobile Ad-hoc Networks. Mobile Network and Applications (MONET), vol 8(5), pp 579–592.
8. Davis CR (2004) A Localized Trust Management Scheme for Ad-hoc Networks. In: Proceedings of the 3rd International Conference on Networking (ICN '04).
9. Eschenauer L, Gligor VD, Baras J (2002) On Trust Establishment in Mobile Ad-hoc Networks. In: Proceedings of the Security Protocols Workshop. Cambridge, U.K.
10. Ghosh T, Pissinou N, Makki K (2005) Towards Designing a Trusted Routing Solution in Mobile Ad-hoc Networks. Mobile Networks and Applications (MONET), vol 10(6), pp 985–995.
11. Ghosh T, Pissinou N, Makki N (2004) Collaborative Trust-based Secure Routing Against Colluding Malicious Nodes in Multi-hop Ad-hoc Networks. In: Proceedings of 29th IEEE Annual Conference on Local Computer Networks (LCN). Tampa, Florida, USA.
12. Gray E, et al. (2002) Trust Propagation in Small Worlds. In: Proceedings of the 1st International Conference on Trust Management. Heraklion, Crete, Greece.
13. He Q, Wu D, Khosla P (2004) SORI: A Secure and Objective Reputation-based Incentive Scheme for Ad-hoc Networks. In: Proceedings of the IEEE Wireless Communications and Networking Conference. Atlanta, Georgia, USA.

14. Jiang T, Baras JS (2004) Ant-based Adaptive Trust Evidence Distribution in MANET. In: Proceedings of the 24th International Conference on Distributed Computing Systems Workshops. Tokyo, Japan.
15. Johnson DB, Maltz DA (1999) The Dynamic Source Routing Protocol for Mobile Ad-hoc Networks", Internet Draft, MANET Working Group, IETF.
16. Josang A (2001) A Logic for Uncertain Probabilities. International Journal of Uncertainty, Fuzziness and Knowledge-based Systems, vol 9(3), pp 279–311.
17. Josang A (1998) A Subjective Metric of Authentication. In: Proceedings of ESORICS: European Symposium on Research in Computer Security.
18. Josang A (1997) Prospectives for Modelling Trust in Information Security. In: Proceedings of Australasian Conference on Information Security and Privacy. Springer, pp 2–13.
19. Kagal L, Finin T, Joshi A (2001) Moving from Security to Distributed Trust in Ubiquitous Computing Environments. IEEE Computer, vol 34(12).
20. Lamsal P (2002) Requirements for Modeling Trust in Ubiquitous Computing and Ad-hoc Networks. Ad-hoc Mobile Wireless Networks- Research Seminar on Telecommunications Software.
21. Li X, Lyu MR, Liu J (2004) A Trust Model Based Routing Protocol for Secure Ad-hoc Networks. In: Proceedings 2004 IEEE Aerospace Conference. Big Sky, Montana, U.S.A.
22. Marti S, Giuli TJ, Lai K, Baker M (2000) Mitigating Routing Misbehavior in Mobile Ad-hoc Networks. In: Proceedings of the 6th annual international conference on Mobile Computing and Networking (MobiCom). Boston, Massachusetts, United States.
23. Michiardi P, Molva R (2002) CORE: A Collaborative Reputation Mechanism to Enforce Node Cooperation in Mobile Ad-hoc Networks. In: Proceedings of the 6th IFIP Communications and Multimedia Security Conference. Portorosz, Slovenia.
24. Ngai ECH, Lyu MR (2004) Trust and Clustering-Based Authentication Services in Mobile Ad-hoc Networks. In: Proceedings of the 2nd International Workshop on Mobile Distributed Computing (MDC'04). Tokyo, Japan.
25. Perkins C, Royer E (1999) Ad-hoc On-Demand Distance Vector Routing. In: Proceedings of the IEEE Workshop on Mobile Computing Systems and Applications.
26. Pirzada AA, McDonald C (2004) Establishing Trust in Pure Ad-hoc Networks. In: 27th Australian Computer Science Conference. Univ. of Otago, Dunedin, New Zealand.
27. Pissinou N, Ghosh T, Makki K (2004) Collaborative Trust Based Secure Routing in Multihop Ad-hoc Networks. In: Proceedings of The Third IFIP-TC6 Networking Conference (Networking '04): Springer Verlag, Series:Lecture Notes in Computer Science, vol 3042, pp1446–1451, Athens, Greece.
28. Theodorakopoulos G, Baras JS (2004) Trust Evaluation in Ad-Hoc Networks. ACM Workshop on Wireless Security (WiSE), Philadelphia, PA, USA.
29. Verma RRS, O'Mahony D, Tewari H (2001) NTM – Progressive Trust Negotiation in Ad-hoc Networks. In: Proceedings of the 1st joint IEI/IEE Symposium on Telecommunications Systems Research. Dublin, Ireland.

30. Yan Z, Zhang P, Virtanen T (2003) Trust Evaluation Based Security Solution in Ad-hoc Networks. In: Proceedings of the Seventh Nordic Workshop on Secure IT Systems.
31. Zeng X, Bagrodia R, Gerla M (1998) Glomosim: A Library for Parallel Simulation of Large-scale Wireless Networks. In: Proceedings of the 12th Workshop on Parallel and Distributed Simulations (PADS). Alberta, Canada.

5 The Concept of Opportunistic Networks and their Research Challenges in Privacy and Security

Leszek Lilien,[1,2] Zille Huma Kamal,[1] Vijay Bhuse,[1] and Ajay Gupta[1]

[1] WiSe (Wireless Sensornet) Lab, Department of Computer Science, Western Michigan University, Kalamazoo, MI 49008, USA

[2] Affiliated with the Center for Education and Research in Information Assurance and Security (CERIAS), Purdue University, West Lafayette, IN 47907, USA

5.1 Introduction

Critical privacy and security challenges confront all researchers and developers working on ever more pervasive computing systems. We belong to this group. We proposed a new paradigm and a new technology of *opportunistic networks* or *oppnets* to enable integration of the diverse communication, computation, sensing, storage and other devices and resources that surround us more and more. We not only find ourselves in their midst but depend on them increasingly as necessities rather than luxuries. As communications and computing systems are becoming more and more pervasive, the related privacy and security challenges become tougher and tougher.

With oppnets, we charted a new direction within the area of computer networks. One of us invented opportunistic *sensor* networks [3]. The idea was later generalized by two of us to general opportunistic networks[1] [31]. To the best of our knowledge we are now the first to scrutinize privacy and security challenges inherent in oppnets.

[1] The name "opportunistic" is used for networks other than our oppnets [41]. In cases known to us, their "opportunism" is quite restricted, e.g., limited to opportunistic communication, realized when devices are within each other's range. In contrast, our oppnets realize opportunistic growth and opportunistic use of resources acquired by this opportunistic growth.

5.1.1 Goal for opportunistic networks

The goal for oppnets is to leverage the wealth of pervasive resources and capabilities that are within their reach. This is often a treasure that remains useless due to "linguistic" barriers. Different devices and systems are either unable speak to each other, or do not even try to communicate. They remain on different wavelengths—sometimes literally, always at least metaphorically.

This occurs despite devices and systems gaining ground in autonomous behavior, self-organization abilities, adaptability to changing environments, or even self-healing when faced with component failures or malicious attacks. It might look somewhat ironic to a person unaware of interoperability challenges that such ever more powerful and intelligent entities are not making equally great strides in talking to each other.

The oppnet goals can be realized by alleviating first of all the communication problems—including bottlenecks and gaps—that are often the root causes of resource shortages (similarly as transportation inadequacies—not the lack of food in the world—are the root causes of famines).

5.1.2 Seed oppnets, helpers, and expanded oppnets

Oppnets and their salient features can be described succinctly as follows. Typically, the nodes of a single network are all deployed together, with the size of the network and locations of its nodes pre-designed (either in a fully "deterministic" fashion, or with a certain degree of randomness, as is the case with ad hoc or mobile networks). In contrast, the size of an oppnet and locations of all but the initial set of its nodes—known as the *seed nodes*—can not be even approximately predicted. This is the category of networks where diverse devices, *not* employed originally as network nodes, are invited to join the seed nodes to become oppnet *helpers.* Helpers perform certain tasks they have been invited (or ordered) to participate in. By integrating helpers into its fold, a *seed oppnet* grows into an *expanded oppnet.*

For example, the seed oppnet shown in Fig. 1 grew into the expanded network shown in Fig. 2 by having admitted the following helpers: (a) a computer network from a nearby college campus, (b) a cellphone infrastructure (represented by the cellphone tower), (c) a satellite, (d) a smart appliance (e.g., a smart refrigerator) providing access to a home network, (e) a microwave relay providing access to a microwave network, (f) a vehicular computer network, connected with wearable computer networks on, and possibly within, the bodies of the occupants of a car.

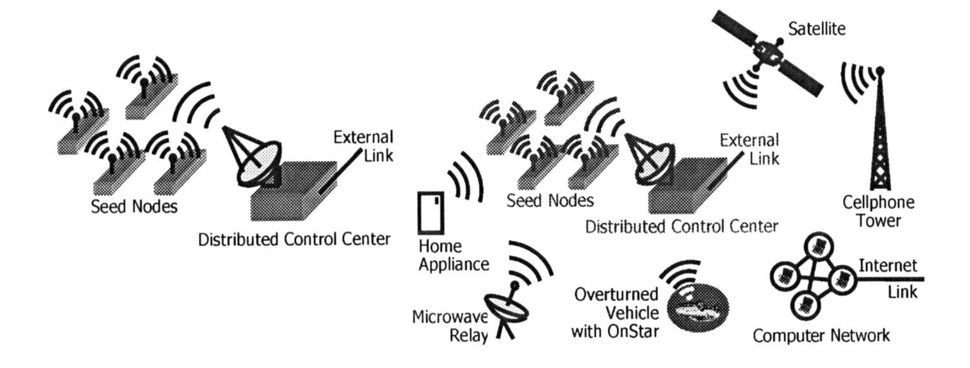

Fig. 1. Seed oppnet.

Fig. 2. Expanded oppnet.

In general, the set of *potential helpers* for oppnets is very broad, incl. communication, computing and sensor systems; wired and wireless; free-standing and embedded. As computing devices continue to become more and more pervasive, the pool of candidates will continue increasing dramatically around us: in infrastructures, buildings, vehicles, appliances, our pockets, etc.

More densely populated areas will have, in general, a denser coverage by potential helpers. As a result, it will be easier to leverage capabilities of an oppnet in more densely populated areas. This is a very desirable natural property, since more resources become available in areas with a possibility of more human victims and more property damage.

With many potential helpers available in an oppnet environment, we need "only" to integrate them in a clever way. We believe that our oppnet paradigm provides a very useful framework—including a conceptual frame of thought—for such integration.

The following scenario illustrates a possible use of an oppnet. A seed oppnet is deployed in a metropolitan area after an earthquake. It finds many potential helpers, and integrates some of them into an expanded oppnet. One of the nodes of the expanded oppnet, a surveillance system, "looks" at a public area scene with many objects. The image is passed to an oppnet node that analyzes it, and recognizes one of the objects as an overturned car (cf. Fig. 2). Another node decides that the license plate of the car should be read. As the oppnet currently includes no image analysis specialist, a helper with such capabilities is found and integrated into the oppnet. It reads the license plate number. The license plate number is used by another newly integrated helper to check in a vehicle database whether the car is equipped with the OnStar™ communication system. If it is, the appropriate OnStar center facility is contacted, becomes a helper, and obtains a connection with the OnStar device in the car. The OnStar device in the car becomes a helper and is asked to contact BANs (body area networks) on and within bodies of car occupants. Each BAN available in the car becomes a helper and reports on the vital signs of its owner. The reports from BANs are analyzed by scheduling nodes that schedule the responder teams to ensure that people in the most serious condition are rescued sooner than the ones that can wait for help longer. (Please note that with the exception of the BAN link that is just a *bit* futuristic—its widespread availability could be measured in years not in decades—all other node and helper capabilities used in the scenario are already quite common.)

5.1.3 Impacts of oppnets

If the researchers, developers, and manufacturers succeed in building oppnets, the payoff will be swift and substantial. Armies of helpers, mobilized by oppnets, will be capable of contributing towards oppnets' objectives at a very low or no cost, the latter especially in emergency situations.

The potential of oppnets in all kinds of emergencies—including man-made and natural disasters—is especially noteworthy. In the past few years we have seen great disasters, such as the 9/11 terrorist attack, the tsunami in the Southeast Asia, and Hurricane Katrina. Casualties and damages are too often compounded by problems faced by the first responders and relief workers. There is a common thread to all these problems: a lack of adequate communication facilities in the disaster areas and beyond. Therefore, providing means of dependable communication in emergencies via oppnets should produce swift and substantial payoffs.

The impact of oppnets on research and development can be significant, especially in the broad and expanding field of pervasive computing. We believe that oppnets are an epitome of pervasive computing. The most critical problems inherent to pervasive computing were very aptly expressed as follows [46]: *Pervasive computing has pervasive problems, not the least of which are interoperability, security and privacy.* Oppnets confront all three enumerated problems head on (though in this chapter we concentrate on the discussion of privacy and security issues). Therefore, work on privacy and security problems in oppnets will be a good test case for attacking the privacy and security problems in pervasive computing.

5.1.4 Chapter contents

The next section describes the basics of oppnet operation. Section 3 describes example oppnet applications and use scenarios. Section 4 presents related work in privacy and security. Section 5 emphasizes the critical significance of privacy challenges in oppnets. Section 6 presents the privacy and security challenges in oppnets, and sketches proposed research directions for solutions to some of these challenges. Finally, Section 7 concludes the paper.

5.2 Opportunistic Networks: Basics of Operation

5.2.1 Seed oppnets and their growth into expanded oppnets

Each opportunistic network grows from a *seed* that is a set of nodes employed together at the time of the initial oppnet deployment. The seed is pre-designed (and can therefore be viewed as a network in its own right). In the extreme case, it can consist of a single node.

The seed grows into a larger network by extending invitations to join the oppnet to foreign devices, node clusters, networks, or other systems which it is able to contact. Any new node that becomes a full-fledged oppnet member, that is a *helper*, may be allowed to invite external nodes. By inviting "free" collaborative nodes, the opportunistic networks can be very competitive economically. The issues that have to be addressed are proper incentives or enforcements so that invited nodes are willing or required to join, and potentially lower credibility of invited collaborators that, in general, can't be fully trusted (at least till they prove themselves). Helpers of an oppnet collaborate on realizing the oppnet's goal. They can be deployed to execute all kinds of tasks even though, in general, they were not designed to become elements of the oppnet that invited them.

5.2.2 Oppnet helpers and oppnet reserve

5.2.2.1 Potential Oppnets Helpers

The set of potential helpers includes even entities not usually thought of as powerful network nodes, both wired and wireless, free-standing and embedded. Even nodes without significant processing, communication, or sensing capabilities, can collectively contribute to processing or communication capabilities of an oppnet in a significant way. After all, any networked PC or embedded processor has *some* useful sensing, processing, or communication capabilities. As examples of minimal useful capabilities, we can consider information about user's presence or absence, her work habits and Internet access patterns collected by her desktop and her PDA; information about user's location collected by his cellphone (even one without GPS can be triangulated); and data about food consumed by user's household collected by a processor embedded in a refrigerator and RFID-equipped food packages and containers.

Before a seed oppnet can grow, it must discover its own set of *potential helpers* available to it. As an example of a discovery, a PC can be discovered by an oppnet once the oppnet identifies a subset of Internet addresses (*IP addresses*) located in its geographical area. Another example of discovery could involve an oppnet node scanning the spectrum for radio signals or beacons, and collecting enough information to be able to contact their senders.

5.2.2.2 Helper Functionalities

It should be noted that, in general, working in the "disaster mode" does not require any new functionalities from the helpers. For example, in case of fire monitoring tasks, the weather sensornet that became a helper can be simply told to stop collecting precipitation data, and use the released resources to increase the sampling rates for temperature and wind direction.

It is possible that more powerful helpers could be reprogrammed on the fly. Also, oppnet nodes might be built with excess general-purpose communication, computation, storage, sensing, and other capabilities useful in case of unforeseen emergencies. For example, excess sensing capabilities could be facilitated by multisensor devices that are becoming cheaper and cheaper as new kinds of sensors are being developed all the time (for example, novel biosensors for detection of anthrax [21]).

Use of helper functionalities can be innovative in at least two ways. First, oppnets are able to exploit *dormant capabilities* of their helpers. For instance, even entities with no obvious sensing capabilities can be used for sensing: (a) a desktop can "sense" its user's presence at the keyboard; (b) a smart refrigerator monitoring opening of its door can "sense" presence of potential victims at home in a disaster area. As another example, the water infrastructure *sensornet* (sensor network) with multisensor capabilities, which is positioned near roads, can be directed to sense vehicular movement, or the lack thereof.

Second, helpers might be used in novel combinations, as illustrated by the scenario from Section 1.2. In the scenario, a complex interaction of many oppnet nodes and helpers starts when a surveillance system, serving as an oppnet node, receives an image of an overturned car.

5.2.2.3 Asking or Ordering Helpers and Oppnet Reserve

Helpers are either invited or ordered to join [33, 32]. In the former case, contacted potential helpers can either volunteer or refuse the invitation. In the latter case, they must accept being conscripted in the spirit of citizens called to arms (or suffer the consequences of going AWOL).

The issue of ordering candidate helpers may seem controversial, and requires addressing. First, it is obvious that any candidate can be asked to join in any situation. Second, any candidate can be ordered to join in life-or-death situations. It is an analogy to citizens being required by law to assist with their property (e.g., vehicles) and their labor in saving lives or critical resources. Third, some candidates can always be ordered to become helpers in emergencies. Such helpers include many kinds of computing and communication systems serving police, firemen, National Guard, and military. Also the federal and local governments can make some of their systems available for any oppnet deployed in an emergency.

The category of systems always available on an order coming from an oppnet includes systems that volunteer—actually, "are volunteered" by their owners. In an obvious analogy to the Army, Air Force, and other Reserves, they all can be named collectively as the *oppnet reserve.* Individually they are *oppnet reservists.* As in the case of the human reserves, volunteers sign up for oppnet reserve for some incentives, be they financial, moral, etc. Once they sign up, they are "trained" for an active duty: facilities assisting oppnets in their discovery and contacting them are installed on them. For example, a standard Oppnet Virtual Machine (OVM) software, matched to their capabilities—either heavy-, medium- or lightweight—is installed on them. (OVM is discussed in [32].) The "training" makes candidates highly prepared for their oppnet duties.

By employing helpers working for free (as volunteers or conscripts), opportunistic networks can be extremely competitive economically in their operation. Full realization of this crucial property requires determining the most appropriate incentives for volunteers and enforcements for conscripts.

5.2.2.4 Preventing Unintended Consequences of Integrating Helpers

Examples of unintended consequences when integrating helpers are disruptions of operations of life-support and life-saving systems, traffic lights, utilities, PTSN and cell phones, the Internet, etc. [32].

To protect critical operations of oppnets and of helpers joining an oppnet, oppnets must obey the following principles:

- Oppnets must not disrupt critical operations of potential helpers. In particular, they must not take over any resources of life-support and life-saving systems.
- For potential helpers running non-critical services, risk evaluation must be performed by an oppnet before they are asked or ordered to join the oppnet. This task may be simplified by potential helpers identifying their own risk levels, according to a standard risk level classification.
- Privacy and security of oppnets and helpers must be assured, especially in the oppnet growth process.

5.2.3 Critical mass for an oppnet and growth limitations

5.2.3.1 Critical Mass

Oppnets can be really effective if they are able to expand their reach enough to reach a certain "critical mass" in terms of size, node locations, and node capabilities. Once this threshold is passed, they are ready to communicate, compute, and sense their physical environment. They can gather data for damage assessment when used in emergencies or disaster recovery. Some sensornets that become helpers—such as sensor nodes embedded in roads, buildings, and bridges—are designed primarily for damage assessment. Other helpers, whether members of sensornets or not, can gather data—legitimately or not—on general public, employees, or other monitored individuals.

5.2.3.2 Growth Limitations

The network stops inviting more nodes when it obtains enough helpers providing sufficient sensing, processing, and communication capabilities (cost/benefit analysis of inviting more nodes might be performed). It should avoid recruiting superfluous nodes that wouldn't help and might reduce performance by using resources just to "gawk." This does not mean that network configuration becomes frozen. As the area affected by the monitored activity (e.g., an earthquake) changes and the required monitoring level in different locations shifts (due, say, to the severity of damage), the oppnet reconfigures dynamically, adapting its scope and its capabilities to its needs (e.g., to the current disaster recovery requirements).

5.3 Example Oppnet Applications and Use Scenarios

5.3.1 Characteristics of oppnet-based applications

Use of oppnets is most beneficial for applications or application classes characterized by the following properties:

- It can start with a seed
- It requires high interoperability
- It uses highly heterogeneous software and hardware components
- It can benefit significantly from leveraging diverse resources of helpers
- It is able to maintain persistent connectivity with helpers once it is established

We are working on a Standard Implementation Framework for oppnets [32] which will facilitate creating oppnet-based applications by providing a standard set of primitives. The primitives for use by application components will, for example, facilitate discovering potential helpers, integrating them, and releasing them when they are not needed any more.

5.3.2 Example oppnet application classes

We can envision numerous applications and application classes that can be facilitated by oppnets. Some of them are described next.

5.3.2.1 Emergency Applications

We see important applications for opportunistic networks in all kinds of emergency situations, for example in hurricane disaster recovery and homeland security emergencies. We believe that they have the potential to significantly improve efficiency and effectiveness of relief and recovery operations. For predictable disasters (like hurricanes or firestorms, whose path can be predicted with some accuracy), seed oppnets can be put into action and their build-up started (or even completed) *before* the disaster, when it is still much easier to locate and invite other nodes and clusters into the oppnet. The first helpers invited by the seed could be the sensornets deployed for structural damage monitoring and assessment, such as the ones embedded in buildings, roads, and bridges.

5.3.2.2 Home/Office Oppnet Applications

Oppnets can benefit home/office applications by utilizing resources within the domestic/office environment to facilitate mundane tasks. Consider contrast between the two scenarios for viewing a visual message on a PDA in a living room. Without an oppnet-based software, PDA has to present the message using the miniscule PDA screen and its substandard speakers. With an oppnet-based software, PDA (now being a single-node seed oppnet) can quickly find helpers: a TV monitor and an audio controller for HiFi speakers available in the living room. PDA can ask these helpers to join, and integrate them into an expanded oppnet. The expanded oppnet, now including 3 nodes (the PDA, the TV monitor, and the audio controller), can present the visual message on high-quality devices.

A similar scenario can be realized in MANETs [38] but with much more programmer's efforts since MANETs do not provide high-level application-oriented primitives to simplify implementation. Only oppnets do [32].

5.3.2.3 Benevolent and Malevolent Oppnet Applications

As most technologies, opportunistic networks can be used to either benefit or harm humans, their artifacts, and technical infrastructure they rely upon. Invited nodes might be "kept in the dark" about the real goals of their host oppnets. Specifically, "good guys" could be cheated by a malevolent oppnet and believe that they will be used to benefit users. Similarly, "bad guys" might be fooled by a benevolent oppnet into believing that they collaborate on objectives to harm users, while in fact they would be closely controlled and participate in realizing positive goals.

On the negative side, home-based opportunistic networks could be the worst violators of individual's privacy, if they are able to exploit PCs, cellphones, computer-connected security cameras, embedded home appliance processors, etc.

5.3.2.4 Predator Oppnets

To counteract malevolent oppnets threats, *predator* networks that feed on all kinds of malevolent networks—including malevolent oppnets—can be created. Using advanced oppnet capabilities and primitives, they can detect malevolent networks, plant spies (oppnet helpers) in them, and use the spies to discover true goals of suspicious networks. Their analysis must be careful, as some of the suspicious networks might actually be benevolent ones, victims of false positives. Conversely, intelligent adversaries can deploy malevolent predator networks that feed on all kinds of benevolent networks, including benevolent oppnets.

5.3.3 Example oppnet application scenarios

We now discuss two example oppnet application scenarios: a benevolent one and a malevolent one. Both rely on some reconfiguration capabilities of non-opportunistic (regular) sensornets.

5.3.3.1 Benevolent Oppnet Scenario —"Citizens Called to Arms"

A seed oppnet is deployed in the area where an earthquake occurred. It is an ad hoc wireless network with nodes much more powerful than in a "typical" ad hoc network (more energy, computing and communication resources, etc.). Once activated, the seed tries to detect any nodes that can help in damage assessment and disaster recovery. It uses any available method for detection of other networks, including radio-based detection (including use of Software Defined Radio and cellphone-based methods), searching for nodes using the IP address range for the affected geographic area, and even AI-based visual detection of some appliances and PCs (after visual detection, the seed still needs to find a network contact for a node to be invited).

The oppnet "calls to arms" the optimal subset of detected and contacted "citizens," inviting all devices, clusters, and entire networks, which are able to help in communicating, computing, sensing, etc. In emergency situations, entities with any sensing capabilities (whether members of sensornets or not), such as cellphones with GPS or desktops equipped with surveillance cameras, can be especially valuable for the oppnet.

Let us suppose that the oppnet is able to contact three independent sensornets in the disaster area, deployed for weather monitoring, water infrastructure control, and public space surveillance. They become helper candidates and are ordered (this is a life-or-death emergency!) to immediately abandon their normal daily functions and start assisting in performing disaster recovery actions. For example, the weather monitoring sensornet can be called upon to sense fires and flooding, the water infrastructure sensornet with multisensor capabilities (and positioned under road surfaces) —to sense vehicular movement and traffic jams, and the public space surveillance sensornet —to automatically search public spaces for images of human victims.

5.3.3.2 Malevolent Oppnet Scenario — "Bad Guys Gang Up"

Suppose that foreign information warriors use agents or people unaware of their goals to create an apparently harmless weather monitoring sensornet.

Only they know that the original sensornet becomes a seed of a malevolent oppnet when activated. The sensornet starts recruiting helpers.

The seed does reveal its true goals to any of its helpers. Instead, it uses a cover of a beneficial application, proclaiming to pursue weather monitoring for research. Actually, this opportunistic sensornet monitors weather but for malicious reasons: it analyzes wind patterns that can contribute to a faster spread of poisonous chemicals. Once the "critical mass" in terms of geographical spread and sensing capabilities is reached, the collected data can be used to make a decision on starting a chemical attack.

5.4 Related Work in Privacy and Security

In this section we discuss briefly some privacy and security solutions proposed in: (a) pervasive computing, (b) ambient networks, (c) grid computing. We also discuss privacy and security solutions based on: (a) trust and reputation in open systems, (b) intrusion detection in ad hoc, mobile, or wireless systems, and (c) honeypots and honeyfarms.

5.4.1 Privacy and security solutions in pervasive computing

Pervasive computing environments require security architecture based on trust rather than just user authentication and access control [25]. Campbell *et al.* [7] looked at the development of several middleware solutions that can support different aspects of security, including authentication, access control, anonymity, and policy management. They also looked at the instantiations of these aspects with diverse mechanisms.

Chen *et al.* [10] described a risk assessment model and proposed an estimator of risk probability that can form the core part of a risk assessment in a ubiquitous computing environment. This estimator is based on a general definition inspired by traditional probability density function approximation, and an implementation by a clustering procedure. To take a distribution of points into account, the authors adopted the Mahalanobis distance for calculating similarities of interactions. They proposed to develop the SECURE framework into which this risk probability estimator is embedded. This risk estimator is feasible and the authors have demonstrated that it fits well within the framework.

Transportation has traditionally been the realm of the machine [13]. Today, as vehicles become increasingly computerized, the authors propose to see this technology moving from under the hood to pervasively connect with passengers, other vehicles and the world. Security and privacy consequences are significant.

Wagealla *et al.* [52] propose a model for trust-based collaboration in ubiquitous computing. The model ensures secure collaboration and interaction between smart devices, by addressing the concerns of security and trust.

Undercoffer *et al.* [47] designed a communications and security infrastructure that goes far in advancing the goal of anywhere-anytime computing. Their work securely enables clients to access and utilize services in heterogeneous networks. It provides a service registration and discovery mechanism implemented through a hierarchy of service management. The system was built upon a simplified PKI that provides for authentication, non-repudiation, anti-playback, and access control. Smartcards were used as secure containers for digital certificates. The system is dependent solely on a base set of access rights for providing distributed trust model. The authors presented the implementation of the system and described the modifications to the design that are required to further enhance distributed trust. They claim that the implementation is applicable to any distributed service infrastructure, whether the infrastructure is wired, mobile, or ad hoc.

Kagal *et al.* [27] used an agent-oriented paradigm to model interactions between computationally enabled entities in pervasive environments. They presented an infrastructure that combined existing authentication features, like SPKI, with notions of policy-driven interaction and distributed trust in order to provide a highly flexible approach for enforcing security policies in pervasive computing environments. They implement the system on a variety of handheld and laptop devices using Bluetooth and 802.11.

5.4.2 Privacy and security solutions in ambient networks

The key problem privacy and security issues in *ambient networks* [43] can be categorized and summarized as follows:

1. Trust establishment and secure agreements
   This includes: (a) a foundation for trust modeling, and (b) security for establishment and execution of general agreements between parties in a dynamic and scalable way.
2. Access security
   This includes: (a) security services at a network edge, e.g., means for a mobile terminal connecting to an access network assuring that it receives the configuration parameters in a secure way, (b) required security services below the IP layer and interfaces to higher-layer control components, and (c) security aspects of ad hoc and multi-hop networks that extend a fixed public network in two cases: (i) where the extension is controlled by the network

operator, and (ii) where individual nodes owned by different parties co-operate to provide a better coverage.

3. Security for mobility and multi-homing
   This includes: (a) security for mobility mechanisms, focusing especially on approaches that do not assume shared authentic-cation infrastructure between all parties, (b) security challenges in mobility mechanisms that optimize movement for groups of nodes simultaneously, (c) security aspects of session mobility, i.e., moving an ongoing session from one device to another, and (d) secure traversal and management of middleboxes, such as firewalls and Network Address Translators (NATs).
4. Special topics
   This includes: (a) group security, e.g., the creation of dynamic, efficient and scalable key management infrastructures for distribution of keys in large groups, and (b) attack resistance dealing with intrusion detection and other methods for protection against threats to availability.

5.4.3 Privacy and security solutions in grid computing

Humphrey and Thompson [20] and Welch *et al.* [53] discuss security-related research in *grid computing*. The Authorization Accounting Architecture Research Group proposes the following high-level requirements [50, 14]:

1. Authorization decisions must be made on the basis of information about the user, the service requested and the operating environment. Information about a user must include extensible attributes as well as the identity. Unknown users must be supported.
2. Identity and attribute information must be passed with integrity, confidentiality, and non-repudiation.
3. Authorization information must be timely (and revocable).
4. Supporting application proxying for users.
5. Supporting ways of expressing trust models between domains.
6. Protocol must support context-sensitive decisions and transactions.
7. Both centralized and distributed administration of authorization information.
8. Separate or combined messages for authentication and authorization.
9. Authorization information should be usable by applications, including accounting and auditing applications.
10. Support negotiation of security parameters between a requestor and a service.

Johnston *et al.* [23] have also written about the special security considerations for grids based on the experience of the NASA Production IPG grid as well as the experience with several DOE collaborators. They considered the threat model and risk reduction in some detail and came up with a security model based on using available grid security services.

5.4.4 Privacy and security solutions based on trust and reputation in open systems

Burnside *et al.* [6] described a resource discovery and communication system designed for security and privacy. All objects in the system, e.g., appliances, wearable gadgets, software agents, and users have associated trusted software proxies that either run on the appliance hardware or on a trusted computer. They described how security and privacy are enforced using two separate protocols: a protocol for secure device-to-proxy communication, and a protocol for secure proxy-to-proxy communication. Using two separate protocols allows running a computationally inexpensive protocol on *thin* devices, and a sophisticated protocol for resource authentication and communication on more powerful devices. The authors designed a device to proxy protocol for lightweight wireless devices, and the proxy-to-proxy protocol which is based on SPKI/SDSI (Simple Public Key Infrastructure / Simple Distributed Security Infrastructure).

The CONFIDANT protocol [5] detects misbehaving nodes by means of observation or reports about several types of attacks. It allows to route around misbehaving nodes and to isolate them from the network. Nodes have a *monitor* for observations, *reputation records* for first-hand and trusted second-hand observations, *trust records* to control trust given to received warnings, and a *path manager* for nodes to adapt their behavior according to reputation.

A collaborative reputation mechanism proposed by Michiardi and Molva [35], has a *watchdog* component. It is complemented by a reputation mechanism that differentiates between *subjective reputation* (observations), *indirect reputation* (positive reports by others), and *functional reputation* (task specific behavior). They are all weighted to derive a combined reputation value that is used to make decisions about cooperation with or a gradual isolation of a node.

Bansal and Baker [2] propose a mechanism that relies exclusively on first-hand observations for *ratings*. If a rating is below the pore-defined *faulty threshold*, the node is added to the *faulty list*. The faulty list is appended to the route request by each node broadcasting the request, and is used as an *avoid list*. A route is rated good or bad depending on whether the next hop is on the faulty list. In addition to the ratings, nodes keep track of the *forwarding balance* with their neighbors, by maintaining a count for each node.

Li *et al.* [29] proposed a new model to quantify trust level of nodes in MANETs. The scheme is distributed and effective without reliance on any central authority. In this scheme, both pre-existing knowledge and direct interaction among nodes in the network can be taken into account as a direct experience for their trust evaluation. To quantify the trust value for direct experiences, the authors defined a new computation function, in which the effect of different direct experience instances can be adjusted individually. To combine own trust value and the recommendation trust value from others, they defined a new trust relationship equation. This scheme deals with the fundamental trust establishment problem and can serve as a building block for higher-level security solutions, such as key management schemes or secure routing protocols.

Venkatraman *et al.* presented [49] an end-to-end data authentication scheme that relies on mutual trust between nodes. The basic strategy is to take advantage of the hierarchical architecture that is implemented for routing purposes. They proposed an authentication scheme that uses TCP at the transport layer and a hierarchical architecture at the IP layer. In this way, the number of encryptions needed is minimized, thereby reducing the computational overheads and resulting in substantial savings, as each node has to maintain keys for fewer nodes.

5.4.5 Privacy and security solutions based on intrusion detection

Mishra *et al.* [36] reviewed many intrusion detection approaches for wireless ad hoc networks.

Nordqvist, Westerdahl and Hansson [39] consider an intrusion detection system for MANETs. Another intrusion detection approach relevant for oppnets comes from the AAFID project [56], in which autonomous agents perform intrusion detection using embedded detectors. An *embedded detector* is an internal software sensor that has added logic for detecting conditions that indicate a specific type of attack or intrusion. Embedded detectors are more resistant to tampering or disabling, because they are a part of the program they monitor. Since they are not executing continuously, they impose a very low CPU overhead. They perform direct monitoring because they have access to the internal data of the programs they monitor. Such data does not have to travel through an external path (a log file, for example) between its generation and its use. This reduces the chances that data will be modified before an intrusion detection component receives them.

Balfanz *et al.* [1] proposed a solution to the problem of secure communication and authentication in ad-hoc wireless networks. The solution provides secure authentication using almost any established public-key-based key exchange protocol, as well as inexpensive hash-based alternatives. In this approach, devices exchange a limited amount of public information over a privileged side channel, which then allows them to complete an authenticated key exchange protocol over the wireless link. This solution does not require a public key infrastructure, is secure against passive attacks on the privileged side channel and all attacks on the wireless link, and directly captures users' intuitions whether they want to talk to a particular, previously unknown device in their physical proximity.

Cross-feature analysis is proposed by Huang, Fan, Lee, and Yu [19] to detect routing anomalies in mobile ad-hoc networks. They explore correlations between features and transform the anomaly detection problem into a set of classification sub-problems. The classifiers are then combined to provide an *anomaly detector.* A sensor facility is required on each node to provide statistics information.

Wireless networks are vulnerable to many identity-based attacks in which a malicious device can use forged MAC addresses to masquerade as a specific client or to create multiple illegitimate identities [12]. A transmitting device can be robustly identified by its *signalprint*, a tuple of signal strength values reported by access points acting as sensors. Apart from MAC addresses or other packet contents, attackers do not have much control regarding the signalprints they produce. By tagging suspicious packets with their corresponding signalprints, the network is able to robustly identify each transmitter independently of packet contents, allowing detection of a large class of identity-based attacks with high probability.

Čapkun *et al.* [9] introduced *integrity regions*, a security primitive that enables integrity protection of messages exchanged between entities that do not hold any mutual authentication material (e.g., public keys or shared secret keys). Integrity regions make use of lightweight ranging techniques and of visual verification within a small physical space. The main application of integrity regions is key establishment. The proposed scheme effectively enables authentication through presence, and therefore protects key establishment from the man-in-the-middle (MITM) attacks. Integrity regions can be efficiently implemented using off-the-shelf components such as ultrasonic ranging hardware.

5.4.6 Privacy and security solutions based on honeypots and honeyfarms

Honeypots and honeyfarms can be considered special types of mechanisms for intrusion detection. A *honeypot* is a decoy whose value lies in being probed, attacked or compromised. It is designed to trap or delay attackers, and gather information about them. Honeypot have resources dedicated to these goals that no other productive value. A honeypot should not see any traffic because it has no legitimate activity. Any interaction with a honeypot is most likely an unauthorized or a malicious activity, and any connection attempts to a honeypot are most likely probes, attacks, or compromises [34]. Honeypot logs can be used to analyze attackers' behaviors and design new defenses.

Honeypots can be categorized with respect to their implementations [22]. A *physical honeypot* is a real machine on the network with its own operating system and address, while a *virtual honeypot* is a Virtual Machine hosted in a physical machine. Virtual honeypots require far less computational and network resources than physical honeypots, and they provide far greater flexibility in emulating various operating systems.

Single honeypots or multiple but independently operated honeypots suffer from a number of limitations, like a limited local view of network attacks, a lack of coordination among honeypots on different networks, inherent security risks involved in honeypot deployment (requiring non-trivial efforts in monitoring and data analysis), and lack of centralized management features. Having a decentralized honeypot presence while providing uniform management in honeypot deployment and operation is a challenging task [22].

A possible solution overcoming the limitations of individual honeypots comes from *honeypot farming*. Instead of deploying large numbers of honeypots in various locations, all honeypots are deployed in a single, consolidated location [44]. This single network of honeypots becomes a *honeyfarm*. Attackers are then redirected to the honeyfarm, regardless of what network they are on or are probing, using *redirectors*. A redirector acts as a proxy transporting an attacker's probes to a honeypot within the honeyfarm, without the attacker ever knowing it. An attacker thinks she is interacting with a victim on a local network, when in reality her attack is transported to the honeyfarm.

5.5 The Critical Significance of Privacy Challenges in Oppnets

The proposed opportunistic network technology is one of possible approaches for moving towards the ultimate goal of pervasive computing. Since huge privacy risks are associated with all pervasive computing approaches, oppnets—being such an approach—must face significant privacy perils.

Pervasiveness must breed privacy threats, as we explain in our 2004 paper [3]:

> *Pervasive devices with inherent communication capabilities might [...] self-organize into huge, opportunistic sensor networks, able to spy anywhere, anytime, on everybody and everything within their midst. [...] Without proper means of detection and neutralization, no one will be able to tell which and how many snoops are active, what data they collect, and who they work for (an advertiser? a nosy neighbor? Big Brother?). Questions such as "Can I trust my refrigerator?" will not be jokes—the refrigerator will be able to snitch on its owner's dietary misbehavior to the owner's doctor.*

We very clearly recognize the crucial issue of privacy in oppnets (as well as in all other pervasive computing approaches). Privacy guarantees are indispensable for realization of the promise of pervasive computing. We strongly believe that without proper privacy protection built into any technology attempting to become pervasive, the public will justifiably revolt against it. Any oppnet solution (or other pervasive computing solution) compromising on privacy protection is doomed to a total failure. Simply, *privacy protection is the "make it or break it" issue for oppnets* and pervasive computing in general.

There is no inherent reason why an oppnet would need to enslave the device asked to help it, exploiting its sensitive resources. There is no inherent reason why the helper device would need to disclose all such resources to the oppnet. In the simplest solution, the candidate helper will keep its private data in a secure vault (e.g., enciphered in its storage) before agreeing to join an oppnet that asked for help. In case of an involuntary conscription (in an emergency situation), the oppnet will allow the candidate helper to save private data in helper's own vault before mustering it.

Other solution we consider will rely on a strict separation of private and public areas within the helper device or network. This will ensure that a benevolent oppnet will never (even when it malfunctions) attempt to capture helper's private data. It will also provide protection against malevolent oppnets that might attack privacy of other devices or networks pretending they need them as their helpers for legitimate needs.

Still other approaches include protecting privacy of helpers and other entities that are under oppnet management or surveillance by, for example, assuring their anonymity or pseudonymity; providing algorithms for detecting malevolent oppnets, which masquerade as benevolent oppnets in order to attack prospective helpers (detection will deny them opportunity to compromise privacy of helpers); and developing methods to protect oppnets against all kinds of privacy attacks, including malicious uses of oppnets for privacy attacks by malicious helpers. The next section describes more privacy solutions.

Some relaxation of the strictest privacy protection standards might be permissible in emergency situation, especially in life-and-death situations. For example, a victim searching for help will probably not object to an oppnet taking over her Body Area Network (BAN), controlling devices on and within her body. We will consider exploring this possibility with a full concern for legal and ethical issues involved. If we do, we will follow two basic assumptions: (1) an entity should give up only as much privacy as is indispensable for becoming a helper for the requesting oppnet; and (2) an entity's privacy disclosure should be proportional to the benefits expected for the entity or to a broader common good. The latter is especially important in emergencies, when the goals like saving a life of one person takes precedence over the comfort of another.

Our earlier work on privacy includes a solution for privacy-preserving data dissemination [30], which we might adapt to improve the oppnet-helper privacy relationships.

Finally, we need to note that privacy (and security) in pervasive computing is a very active investigation area. We can use many other privacy solutions conceived by other researchers working on networks and in the area of pervasive computing.

5.6 Privacy and Security Challenges in Oppnets

One of the main sources of security and privacy threats in oppnets is the fact that even a perfect helper authentication, performed before helpers join oppnets, will not guarantee excluding malicious devices from oppnets. The reason is that even a perfect helper authentication will not preclude abuses of authorizations by insiders. In general, oppnets have to use two lines of defense: (a) *preventive defense*, by blocking malicious helpers from joining them (e.g., by best authentication possible), and (b) *reactive defense*, by detection of malicious devices only after they join them, and their notorious behavior is detected (e.g., by intrusion detection systems).

The most important security and privacy challenges for opportunistic networks, discussed in turn in the following subsections, are:

1. Increasing trust and providing secure routing
2. Helper privacy and oppnet privacy
3. Protecting data privacy
4. Ensuring data integrity
5. Authentication of oppnet nodes and helpers

6. Dealing with specific most dangerous attacks
7. Intrusion detection
8. Honeypots and honeyfarms

Figure 3 displays a general security scheme for oppnets. In the absence of a highly trustworthy authentication mechanism all five steps marked by outgoing arrows from the adder circle are mandatory.

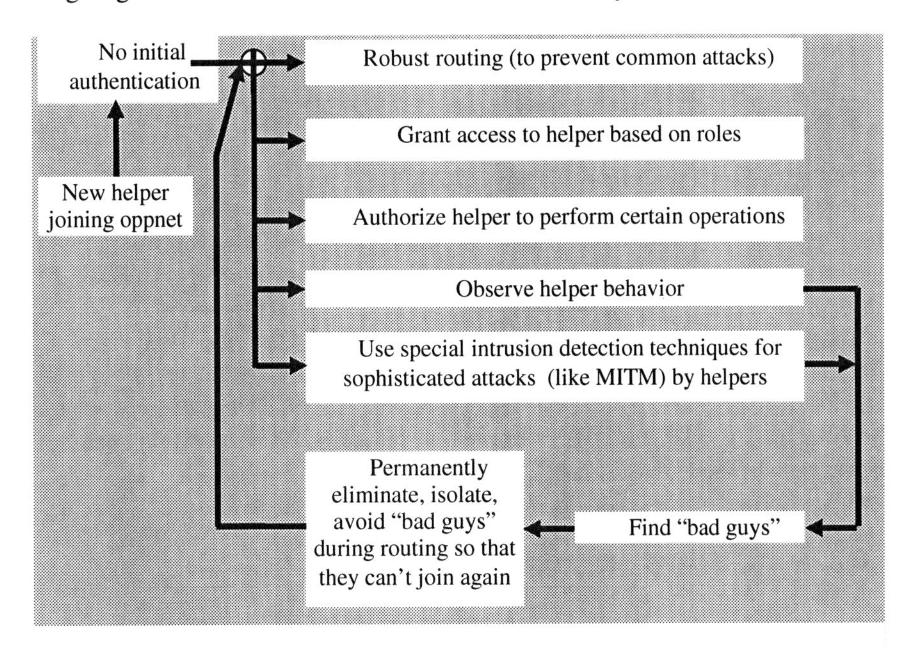

Fig. 3. The general security scheme for oppnets.

5.6.1 Increasing trust and providing secure routing

A list of "more trusted" devices, based on direct experience and second-hand reputation, can be maintained by an oppnet. For example, an oppnet can trust more oppnet reservists, or devices owned by certain institutions, such as devices at police stations, government offices, hospitals, public libraries, universities or reputable companies. Once a list of trusted devices is made, these devices can be used for more critical tasks that should not be entrusted to unknown devices or, even worse, distrusted devices. (A 'black list' of distrusted entities can be maintained as well.)

Secure routing can use both lists. Selecting a route that passes through only trusted devices (or as many trusted devices as possible) is challenging. Numerous papers have been written on individual ad hoc routing protocols. Hu and Perrig wrote a survey of secure wireless ad hoc routing [17].

The secure wireless ad hoc routing protocol that seems most relevant to oppnets is Ariadne [18]. It is an on-demand protocol that works in the presence of compromised nodes. Ariadne uses symmetric cryptography. It authenticates routing messages using one of the three schemes:

- Shared secrets between each pair of nodes
- Shared secrets between communicating nodes combined with broadcast authentication
- Digital signatures

General solutions proposed for securing routing protocols in wireless or ad hoc networks or the Internet cannot be used directly in oppnets because of their special characteristics. Among others, oppnets are highly heterogeneous, with different processing abilities, power sources, modes of transmission, etc.

Trusted devices powered by batteries should be used sparingly to increase their lifetime, and in this way optimizing oppnet connectivity and thus routing. Having adequate battery power might be easier in oppnets than in other systems since oppnets can rely on harvesting needed resources via their growth.

5.6.2 Helper privacy and oppnet privacy

Some approaches for assuring privacy were mentioned in Section 5. More details for some of these solutions and other solutions are presented here.

5.6.2.1 Helper Privacy

To be accepted, oppnets must assure privacy of helpers. A fear of having its privacy violated can prevent candidate helpers invited by an oppnet from joining it, or can cause reluctance (a passive or an active resistance) of candidate helpers ordered by an oppnet to join.

The first line of privacy defense for a helper are its access controls (authentication and authorization) and its intrusion prevention (using security primitives, relying on trust, using secure routing etc.). Intrusion detection should be the second line of privacy defense for helpers, helping when prevention fails or cannot be used due to its inefficiency. Elimination or isolation of bad entities from oppnets via intrusion detection is very important for benevolent nodes. The problem of enforcing access control and performing real-time intrusion detection for oppnets are more difficult than for the Internet, wireless networks, or ad hoc networks in general because of the highly heterogeneous nature of oppnet components and the spontaneous manner in which oppnets are formed.

We investigate three helper privacy approaches: (1) extending initiator anonymity protocols ([15, 42]); (2) providing responder anonymity and anonymous data transfer via proxy techniques [42]; and (3) use of *active certificates* [4] to safeguard sensitive information or resources on helper nodes from software agents sent by an oppnet.

5.6.2.2 Oppnet Privacy

Guarding an oppnet against privacy violations by a helper or by another oppnet node is equally important. Malicious helpers can join an oppnet with the purpose of violating its privacy. Since it is very difficult to uncover the motives of any helper invited or ordered by an oppnet to join, the only way to find bad helpers may be by intrusion detection.

We investigate three helper privacy approaches: (1) a solution based on automatic trust negotiation [55]; (2) using Semantic Web technologies to manage trust [8]; and (3) automatic enforcement of privacy policies, described by the Semantic Web Rule Language (SWRL) [45]. We also started investigation of a Semantic Web framework with an OWL-based ontology [11], and Rei [40, 24] and KAoS [26] policy languages to move towards context-aware policies for oppnets.

5.6.3 Protecting data privacy

Privacy of messages in oppnets is our next concern, considered separately from the privacy issues considered above.

5.6.3.1 Multicast from the Controller

Many controller messages are intended for only a few selected nodes in the oppnet and require privacy. The lack of a shared secret or a key between the controller and the intended recipients makes the problem of providing data privacy difficult. If there is a shared secret key (for the symmetric key cryptography encryption) between the controller and intended recipients, a capture of even a single device leads to the failure of the whole scheme. The capture might be more probable in crisis situations when providing physical protection is even more difficult.

5.6.3.2 Messages from Nodes to the Controller

Many messages from oppnet nodes to its controller also require privacy. Encryption is a standard way of providing such data privacy. Asymmetric key cryptography (or a public key cryptography using PKI) can be used to protect privacy of these messages.

A malicious device can pose as an oppnet controller and distribute its own public key. To prevent distribution of such a forged public key, the legitimate controller needs a secure mechanism to broadcast a public key to oppnet nodes, including candidate helpers and integrated helpers.

Messages in oppnets can be sent from one device to another device (peer to peer) in a wide area or locally. The latter case includes an intra-cluster communication among devices in a neighborhood. A local cluster head (a trusted device) can again use public key cryptography in communicating with its neighbors. A malicious device posing as a cluster head must be prevented from distributing its own forged public key.

5.6.4 Ensuring data integrity

Data integrity is a part of any secure communication. Digital signatures can be used to guard integrity of messages. They are often too expensive computationally for *thin* devices (like cellphones, PDAs, etc.), typically running on a limited battery power. Lightweight alternatives should be devised to guarantee integrity of data packets.

Message size may vary when it travels through an oppnet. Suppose that a message is sent from a cellphone to a base station through a PC connected to the Internet. The size of the packets traveling from the cellphone to the PC will be different from the size of the packets when they travels from the PC to the base station. If packet fragmentation and aggregation cannot be performed securely, the end-to-end security mechanisms assuring data integrity could fail.

5.6.5 Authentication of oppnet nodes and helpers

Delivering secret keys securely to all non-malicious devices (and only to non-malicious devices) is very difficult in ad hoc oppnet environments. Hence, relying alone on cryptography-based authentication mechanisms (such as Kerberos) is not sufficient. We need to deal with a host of sophisticated attacks, such as MITM, packet dropping, ID spoofing (masquerading), and distributed DoS attacks—all significant and potentially disabling threats for oppnets.

We investigate two helper privacy approaches: (1) a solution integrating existing techniques of authentication, authorization and accounting (AAA) ([37, 1]) to provide authentication of nodes in oppnets; (2) use of Identity Based Encryption (IBE) [16] for creating and storing pre-shared secrets, public keys and revocation lists.

5.6.6 Proposed solutions for dealing with specific attacks

The most dangerous attacks on oppnets and their effects can be described briefly as follows:

1. *MITM attacks*: Suppose that a malicious device is on the path connecting a victim and the rescue team. When the victim sends a help request message to the rescuers, the malicious device might capture it and maliciously inform the victim that help is on the way. It could also tamper with messages sent by the rescuers.
2. *Packet dropping*: The malicious device in the above scenario might drop some or all packets sent between the victim and the rescue team. It might capture packets at random, or forward packets containing insignificant information and drop packets containing critical information.
3. *DoS attacks by malicious devices*: False requests for help can be generated by malicious devices. They will keep the rescue team busy and unavailable for real emergencies.
4. *DoS attacks on weak links*: DoS attacks may target a "weak" device, such as a cellphone, that is critical to oppnet operation (e.g., if it is the device that connects two clusters of users). The battery of such a cellphone is a very precious resource and should be used sparingly till an alternative inter-cluster connection is found. Attacks to exhaust battery power can occur. Some DoS attacks may target only critical weak devices.
5. *ID spoofing*: Mapping some node properties (like location of a node) into node ID by a controller can be dangerous. A masquerading malicious device can generate requests with multiple IDs, resulting in many false alarms for the rescue team. Services that need authentication can be misused if IDs can be spoofed. A device capable of spoofing ID of a trusted node or a node with critical functions can pose many kinds of attacks.
6. *Helpers masquerading as oppnet members*: Helper nodes that masquerade as oppnet members can attack the oppnet not only individually, but can form a gang for attacking the oppnet.

Research directions or initial solutions, explored by us to prevent the above attacks, can be sketched as follows:

1. *Solution directions for MITM attacks:* A person in need can send redundant messages to the controller through multiple neighbors. This will increase the chances that least one of the multiple message copies will reach the controller, even if there are attackers on some paths. So, redundancy of routes can be exploited to avoid the MITM attackers. Use of integrity regions [9] is another solution to be investigated for preventing MITM attacks.
2. *Solution directions for packet dropping:* The above idea of sending redundant messages via multiple neighbors may work if no packet-dropping adversary is situated on at least one path. Again, redundancy of routes can be exploited to avoid attackers.
3. *Solution directions for DoS attacks by malicious devices:* Upper limit can be placed on the number of requests any device can generate. Thus, it will limit the number of times any device can send a false help request. In addition, the rescue team can attempt contacting the requester to confirm an emergency request.
   Other solutions under investigation include: (1) integrating a trust evaluation technique [29] and locality driven key management architecture [54]; and (2) a solution based on tagging packets with signalprints [12] and using appropriate matching rules to detect DoS attacks based on MAC address spoofing.
4. *Solution directions for DoS attacks on weak links*: Identification of weak devices, their strengthening (e.g., providing backups for them), or minimizing their workload can counteract such attacks and maintain connectivity in oppnets.
5 *Solution directions for ID spoofing*: Although it is difficult to guarantee that malicious nodes will not join an oppnet, oppnet nodes can watch their neighbors for possible attempts of ID spoofing. The SAVE protocol [28] can provide routers with information needed for source address validation. This protocol needs to be modified to suit the heterogeneous nature of oppnets.
6 *Solution directions for helpers masquerading as oppnet members*: Helpers should be required to at least authenticate/authorize themselves before they can start inviting/ordering other nodes to join the network. We investigate a solution based on signalprints [12], which are highly correlated to physical node locations, and can detect malicious nodes lying about their MAC addresses.

5.6.7 Intrusion detection

Malicious devices or malicious networks can join an oppnet if an initial authentication mechanism is not adequate. There is a need to detect and isolate malicious nodes, clusters, or networks. Securely *distributing* information about malicious entities in the presence of malicious entities is a challenge. If shared securely, this reputation information can be used by all oppnet nodes to protect themselves from attackers. Even if this information can be distributed securely, avoiding the suspected entities while maintaining connectivity is a challenge.

We are investigating requirements for efficient algorithms and protocols for intrusion detection in oppnets, based on existing solutions for MANETs [39]. The characteristics of oppnets make real-time intrusion detection and response in them even more challenging than in other types of networks.

5.6.8 Honeypots and honeyfarms

Design of low-cost honeypots for oppnets is challenging because physical security of honeypots cannot be guaranteed for the entire lifetime of an oppnet. Observations from honeypots cannot be trusted unless secure channels of communication are established. Attackers masquerading as honeypots or posing DoS attacks on honeypots are examples of problems that need to be solved.

We are investigating a hybrid honeyfarm architecture for oppnets that integrates the high-interaction technologies of Collapsar honeyfarm [22] and Potemkin honeyfarm [51], providing both *centralized* management and *decentralized* honeypot presence. The resulting system can be made scalable and efficient, using late binding of resources, flash cloning, and redirectors.

5.7 Conclusions

This chapter describes the concept of *opportunistic networks* (*oppnets*), and presents the related research challenges in privacy and security.

Oppnets constitute a newly identified category of computer networks. When deployed, oppnets attempt to detect candidate helper systems existing in their relative vicinity—ranging from sensing and monitoring, to computing and communication systems—and integrate them under their own control. When such a candidate is detected by an oppnet, the oppnet evaluates the benefits that it could realize if the candidate joins it. If the evaluation is positive the oppnet invites the candidate to become its helper. In this manner, an oppnet can grow from a small seed into a large network with vast sensing, communication, and computation capabilities.

Oppnets will facilitate many applications. As an example, they can help building an integrated network called for in various critical or emergency situations [48]. Oppnets can be used to enable connectivity in an area where any existing communication or information infrastructure has been fractured or partially destroyed. Oppnets will integrate various systems that were not designed to work together. The integration will enhance the flow of information that, for example, can assist in rescue and recovery efforts for devastated areas, or can provide more data on phenomena that are just developing, such as wildfires or flash torrents.

Answering to the identified privacy and security challenges in oppnets will contribute to advancing knowledge and understanding not only for the opportunistic networks, but will simultaneously advance the state of the art of computer privacy and security in ad hoc and in general-purpose computer networks.

We continue working on a number of the identified challenges, continuing our investigation of privacy and security in oppnets. The planned prototype opportunistic network will provide a proof of concept for our solutions, as well as stimulation and feedback necessary for fine-tuning the proposed solutions.

Acknowledgements

This work was supported in part by the National Science Foundation under Grant IIS-0242840, and in part by the U.S. Department of Commerce under Grant BS123456.

The authors would also like to acknowledge Western Michigan University for its support and its contributions to the WiSe (Wireless Sensornet) Laboratory, Computational Science Center and Information Technology and Image Analysis (ITIA) Center.

L. Lilien, a co-PI on the NSF grant providing a partial support for this research, would like to thank Professor Bharat Bhargava from Purdue University, the PI for this grant.

L. Lilien would like to thank the participants of the *International Workshop on Research Challenges in Security and Privacy for Mobile and Wireless Networks* (*WSPWN 2006*) for their helpful comments and feedback on oppnets. In particular, he would like to thank Mr. Hien Nguyen, a Ph.D. student at the Florida International University, for a fruitful discussion that resulted in crystallizing the idea of the oppnet reserve.

L. Lilien also expresses his thanks to the following students of his advanced computer security course for their contributions to the following research projects: (a) contributors to helper privacy and oppnet privacy: N. Bhargava, T. Goodman, V. Kalvala, H.R. Ravi, R. Rekala, A. Rudra, V. Talati, and Y. Yoder ; (b) contributors to authentication of oppnet nodes and helpers: V.V. Krishna, P.E. Miller, and A.K. Yedugani; (c) contributors to dealing with specific attacks: S. Chittineni, N. Jawanda, D. Koka, S. Pulimamidi, and H. Singh; and (d) contributors to intrusion detection, honeypots and honeyfarms: R. Dondati and S. Mapakshi.

Any opinions, finding, conclusions or recommendation expressed in the paper are those of the authors and do not necessarily reflect the views of the funding agencies or institutions.

References

1. D. Balfanz, D. K. Smetters, P. Stewart and H. C. Wong, "Talking To Strangers: Authentication in Ad-Hoc Wireless Networks," *Symposium on Network and Distributed Systems Security (NDSS '02)*, San Diego, CA, Feb. 2002.
2. S. Bansal and M. Baker, "Observation based cooperation enforcement in ad hoc networks," CoRR, July 2003. Available at http://www.informatik.uni-trier.de/~ley/db/journals/corr/corr0307.html#cs-NI-0307012.
3. B. Bhargava, L. Lilien, A. Rosenthal, and M. Winslett, "Pervasive Trust," *IEEE Intelligent Systems*, vol. 19(5), Sep./Oct.2004, pp. 74–77.
4. N. Borisov, "Active Certificates: A Framework for Delegation," M.S. Dissertation, University of California, Berkeley, 2002.
5. S. Buchegger and J. Le Boudec, "Performance Analysis of the CONFIDANT Protocol: Cooperation Of Nodes — Fairness in Dynamic Ad-hoc Networks," *13 IEEE/ACM Symposium on Mobile Ad Hoc Networking and Computing* (*MobiHoc 2002*), Lausanne, Switzerland, June 2002.
6. M. Burnside, D. Clarke, Mills, A. Maywah, S. Devadas, R. Rivest, "Proxy-Based Security Protocols in Networked Mobile Devices", 17th *ACM Symp. on Applied Computing* (*SAC'02)*, Madrid, Spain, March 2002, pp. 265–272.
7. R. Campbell, J. Al-Muhtadi, P. Naldurg, G. Sampemane and M.D. Mickunas, "Towards Security and Privacy for Pervasive Computing," *IEEE Computer*, vol. 34 (12), Dec. 2001, pp. 154–157.
8. O. Can, and M. Unalir, "Distributed Policy Management in Semantic Web," Dept. of Computer Engineering, Ege University Bornova, Izmir, Turkey, 2006.
9. S. Čapkun and M. Cagalj, "Integrity Regions: authentication through presence in wireless networks," *5th ACM Workshop on Wireless Security* (*WiSe'06*), Los Angeles, CA, Sep. 2006, pp. 1–10.

10. Y. Chen, C. Jensen, E. Gray, V. Cahill, J. Seigneur, "A General Risk Assessment of Security in Pervasive Computing," Technical Report TCD-CS-2003-45, Dept. of Computer Science, Trinity College, Dublin, Ireland, Nov. 2003.
11. A. Dersingh, R. Liscano, and A. Jost, "Using Semantic Policies for Ad Hoc Coalition Access Control," *International Workshop on Ubiquitous Access Control* (*IWUAC'06*), San Jose, CA, 2006.
12. D.B. Faria and D.R. Cheriton, "Detecting Identity Based Attacks in Wireless Networks Using Signalprints," *5th ACM Workshop on Wireless Security (WiSe'06)*, Los Angeles, CA, Sep. 2006.
13. K. Farkas, J. Heidemann, and L. Iftode, "Intelligent Transportation and Pervasive Computing," *IEEE Pervasive Computing*, vol. 5 (4), Oct. 2006, pp. 18–19.
14. S. Farrell, J. Vollbrecht, P. Calhoun, L. Gommans, G. Gross, B. DB Bruijn, C. DB Laat, M. Holdrege, and D. Spence, "AAA Authorization Requirements," RFC 2906, The Internet Society, Aug. 2000. Available at: http://www.faqs.org/rfcs/ rfc2906.html.
15. I. Goldberg and D. Wagner, "Taz Servers and the Rewebber Network: Enabling Anonymous Publishing on the World Wide Web," *First Monday*, 1998.
16. K. Hoeper and G. Gong, "Bootstrapping Security in Mobile Ad Hoc Networks Using Identity-Based Schemes with Key Revocation," Technical Report CACR 2006-04, Centre for Applied Cryptographic Research, Waterloo, Canada, 2006.
17. Y.-C. Hu and A. Perrig, "A Survey of Secure Wireless Ad Hoc Routing," *IEEE Security & Privacy, Special Issue on Making Wireless Work*, Vol. 2(3), May/June 2004, pp.28–39.
18. Y.-C. Hu, A. Perrig, and D.B. Johnson, "Ariadne: A Secure On-Demand Routing Protocol for Ad Hoc Networks," *8th Ann. Intl. Conf. Mobile Computing and Networking* (*MobiCom 2002*), Atlanta, Georgia, Sep. 2002, pp. 12–23.
19. Y. Huang, W. Fan, W. Lee, and P. S. Yu, "Cross-feature analysis for detecting ad-hoc routing anomalies," *23rd International Conference on Distributed Computing Systems* (*ICDCS 2003*), Providence, RI, May 2003, pp. 478–487.
20. M. Humphrey and M. Thompson, "Security Implications of Typical Grid Computing Usage Scenarios," *10th IEEE International Symposium on High Performance Distributed Computing*, San Francisco, CA, Aug. 2001, pp. 95–103.
21. H. Inerowicz, S. Howell, F. Regnier, and R. Reifenberger, "Protein Microarray Fabrication for Immunosensing," *224th American Chemical Society (ACS) National Meeting*, Aug. 2002.
22. X. Jiang and D. Xu, "Collapsar: a VM-based Architecture for Network Attack Detection Center," *13th Usenix Security Symposium*, San Diego, CA, Aug. 2004. Available at: www.ise.gmu.edu/~xjiang/pubs/JPDC06.pdf
23. W. E. Johnston, K. Jackson, and S. Talwar, "Security Considerations for Computational and Data Grids," *10th IEEE Symposium on High Performance Distributed Computing*, San Francisco, CA, Aug. 2001.

24. L. Kagal and T. Berners-Lee, "Rein: Where Policies Meet Rules in the Semantic Web," Technical Report, MIT, 2005.
25. L. Kagal, T. Finin, and A. Joshi, "Trust-Based Security in Pervasive Computing Environments," *IEEE Computer*, vol. 34 (12), Dec. 2001, pp. 154–157.
26. L. Kagal, M. Paolucci, N. Srinivasan, G. Denker, T. Finin, and K. Sycara, "Authorization and Privacy for Semantic Web Services," *First International Semantic Web Services Symposium, AAAI 2004 Spring Symposium*, March 2004.
27. L. Kagal, J. Undercoffer, F. Perich, A. Joshi, T. Finin, and Y. Yesha, "Vigil: Providing Trust for Enhanced Security in Pervasive Systems," Dept. of CSEE, University of Maryland Baltimore County, August 20021. Available at: http://ebiquity.umbc.edu/paper/html/id/54/Vigil-Providing-Trust-for-Enhanced-Security-in-Pervasive-Systems
28. J. Li, J. Mirkovic, M. Wang, P. Reiher, and L. Zhang. "SAVE: Source Address Validity Enforcement Protocol," UCLA Technical Report 01-0004, Los Angeles, CA, 2001.
29. X. Li, J. Slay, and S. Yu, "Evaluating Trust in Mobile Ad hoc Networks," *The Workshop of International Conference on Computational Intelligence and Security*, Dec. 2005, Xi'an, China. Available at: http://esm.cis.unisa.edu.au/new_esml/resources/publications/evaluating%20trust%20in%20mobile%20ad-hoc%20networks.pdf
30. L. Lilien and B. Bhargava, "A Scheme for Privacy-preserving Data Dissemination," *IEEE Transactions on Systems, Man and Cybernetics Cybernetics, Part A: Systems and Humans*, Vol. 36(3), May 2006, pp. 503–506.
31. L. Lilien and A. Gupta, Personal Communication, Department of Computer Science, Western Michigan University, Kalamazoo, MI, Dec. 2005.
32. L. Lilien, A. Gupta, and Z. Yang, "Opportunistic Networks and Their Emergency Applications and Standard Implementation Framework," submitted for publication.
33. L. Lilien, Z. H. Kamal, and A. Gupta, :Opportunistic Networks: Research Challenges in Specializing the P2P Paradigm," *3rd International Workshop on P2P Data Management, Security and Trust* (*PDMST'06*), Kraków, Poland, Sep. 2006.
34. M. Locasto, J. Parekh, A. Keromytis, S. Stolfo, "Towards Collaborative Security and P2P Intrusion Detection," *2005 IEEE Workshop on Information Assurance and Security*, June 2005. Available at: http://www1.cs.columbia.edu/ids/publications/locasto2005iaw.pdf
35. P. Michiardi and R. Molva, "CORE: A collaborative reputation mechanism to enforce node cooperation in mobile ad hoc networks," *Sixth IFIP Conference on Security Communications, and Multimedia* (*CMS 2002*), Portorož, Slovenia, Sep. 2002.
36. A. Mishra, K. Nadkarni, A. Patcha, "Intrusion Detection in Wireless Ad Hoc Networks", *IEEE Wireless Communications*, Vol. 11(1), Feb. 2004, pp. 48–60.
37. H. Moustafa, G. Burdon, and Y. Gourhant, "Authentication, Authorization and Accounting (AAA) in Hybrid Ad hoc Hotspot's Environments," *4th*

International Workshop on Wireless Mobile Applications and Services on WLAN Hotspots (*WMASH 2006*), Los Angeles, CA, Sep. 2006.

38. M. Mutka, Personal Communication, Department of Computer Science and Engineering, Michigan State University, East Lansing, MI, Dec. 2006.
39. D. Nordqvist, L. Westerdahl and A. Hansson, "Intrusion Detection System and Response for Mobile Ad hoc Networks," FOI-R 1683, Command and Control Systems, User Report, July 2005.
40. D. Olmedilla, "Security and Privacy on the Semantic Web," in: M. Petkovic and W. Jonker (editors), *Security, Privacy and Trust in Modern Data Management*, Springer, 2006.
41. L. Pelusi, A. Passarella, and M. Conti, "Opportunistic Networking: Data Forwarding in Disconnected Mobile Ad Hoc Networks," *IEEE Communications*, Vol. 44(11), Nov. 2006, pp. 134–141.
42. A. Pfitzmann and M. Waidner, "Networks Without User Observability — Design Options," *Eurocrypt '85, Workshop on the Theory and Application of of Cryptographic Techniques*, Linz, Austria, April 1985, pp. 245–253.
43. G. Selander *et al.*, "Ambient Network Intermediate Security Architecture," Deliverable 7.1, v. 3.2, Ambient Networks Project, Sixth Framework Programme, European Union, Feb. 2005. Available at: www.ambient-networks.org/phase1web/ publications/D7-1_PU.pdf.
44. L. Spitzner, "Definitions and Value of Honeypots", GovernmentSecurity.org, May 2002. Available at: http://www.trackinghackers.com/papers/honeypots.html
45. "SWRL: A Semantic Web Rule Language Combining OWL and RuleML," The World Wide Web Consortium (W3C), May 2004. Available at: http://www.w3.org/Submissions/SWRL/
46. P. Thibodeau, "Pervasive computing has pervasive problems," *ComputerWorld*, Vol. 36(41), Oct. 7, 2002.
47. J. Undercoffer, F. Perich, A. Cedillnik, L. Kagal, A. Joshi, "A Secure Infrastructure for Service Discovery and Access in Pervasive Computing," Technical Report, TR-CS-01-12, Dept. of CSEE, University of Maryland Baltimore County, 2001. Available at: http://citeseer.ist.psu.edu/cedilnik01secure.html.
48. U.S. Government Printing Office via GPO Access, "Combating Terrorism: Assessing the Threat of a Biological Weapons Attack." Last accessed on December 15, 2005. Available at: http://www.armscontrolcenter.org/cbw/resources/hearings/snsvair_20011012_combating_terrorism_assessing_biological_weapons_attack.htm
49. L. Venkatraman and D. Agrawal, "A novel authentication scheme for ad hoc networks", *Wireless Communications and Networking Conference* (*WCNC 2000*), Vol. 3, Chicago, IL, Sep. 2000, pp. 1268–1273.
50. J. Vollbrecht, P. Calhoun, S. Farrell, L. Gommans, G. Gross, B. de Bruijn, C. de Laat, M. Holdrege, D. Spence, "RFC 2905 - AAA Authorization Application Examples", Network Working Group, The Internet Society, Aug. 2000. Available at: www.faqs.org/rfcs/rfc2905.html.
51. M. Vrable, J. Ma, J. Chen, D. Moore, E. Vandekieft, A. C. Snoeren, G. M. Voelker, and S. Savage, "Scalability, Fidelity and Containment in

Potemkin Virtual Honeyfarm," *ACM Symposium on Operating System Principles (SOSP'05)*, Brighton, UK, Oct. 2005.
52. W. Wagealla, C. English, S. Terzis, and P. Nixon, "A Trust-based Collaboration Model for Ubiquitous Computing," *Ubicomp2002 Security Workshop*, Goteborg, Sweden, Sept./Oct. 2002.
53. V. Welch, F. Siebenlist, I. Foster, J. Bresnahan, K. Czajkowski, J. Gawor, C. Kesselman, S. Meder, L. Pearlman, and S. Tuecke, "Security for Grid Services," *Intl. Symp. on High Performance Distributed Computing*, Seattle, WA, June 2003, pp. 48-57. Available at: citeseer.ist.psu.edu/welch03security.html.
54. G. Xu and L. Iftode, "Locality Driven Key Management Architecture for Mobile Ad hoc Networks," *IEEE International Conference on Mobile Ad-hoc and Sensor Systems*, Fort Lauderdale, FL, Oct. 2004.
55. T. Yu, M. Winslett, and K. E. Seamons, "Supporting Structured Credentials and Sensitive Policies through Interoperable Strategies for Automated Trust Negotiation," *ACM Transactions on Information and System Security (TISSEC)*, 6(1), Feb. 2003.
56. D. Zamboni, "Using Internal Sensors for Computer Intrusion Detection," CERIAS Technical Report 2001-42, CERIAS, Purdue University, West Lafayette, IN, Aug. 2001.

6 On Performance Cost of On-demand Anonymous Routing Protocols in Mobile Ad Hoc Networks

Jiejun Kong[1], Jun Liu[2], Xiaoyan Hong[2], Dapeng Wu[3], and Mario Gerla[4]

[1] Jiejun Kong is currently with Scalable Network Technologies, Inc., 6701 Center Drive West, Suite 520, Los Angeles, CA 90045.
[2] Jun Liu and Xiaoyan Hong are with the Department of Computer Science, University of Alabama, Tuscaloosa, AL 35487.
[3] Dapeng Wu is with the Department of Electric and Computer Engineering, University of Florida, Gainesville, FL 32611.
[4] Mario Gerla is with the Department of Computer Science, University of California, Los Angeles, CA 90095.

6.1 Introduction

A mobile ad hoc network (MANET) can establish an instant communication structure for many time-critical and mission-critical applications. Nevertheless, the intrinsic characteristics of ad hoc networks, such as wireless transmission and node mobility, make it very vulnerable to security threats. Many security protocol suites have been proposed to protect wireless communications, however, they do not consider anonymity protection and leave identity information freely available to nearby *passive* eavesdroppers. The goal of passive attacks is very different from those of other attacks on routing, such as route disruption or "denial-of-service" attacks. In fact, the passive enemy will avoid such aggressive schemes, in the attempt to be as "invisible" as possible, until it traces, locates, and then physically destroys legitimate assets [29, 51]. Consider for example a battlefield scenario with ad hoc, multi-hop wireless communications support. The adversary could deploy reconnaissance and surveillance sensor networks in the battlefield and maintain communications among them. Via intercepted wireless transmissions, they could infer the location, movement, number of participants, and even the goals of our task forces. Anonymity and location privacy guarantees for our ad hoc networks are critical, otherwise the entire mission may be compromised. This poses challenging constraints on routing and data forwarding.

6.1.1 Mobile sensor networks

Recent advances in manufacturing technologies have enabled the physical realization of small, light-weight, low-power, and low-cost micro air vehicles (MAVs) [21,22]. These MAVs refer to a new breed of unmanned air vehicles (UAVs) or aerial robots that are significantly smaller than currently available UAVs. Figure 6.1(a) illustrates the WASP MAV recently tested by DARPA. It is a 32 cm "flying wing" made of a plastic lithium-ion battery material that provides both electrical power and wing structure. The wing utilizes synthetic battery materials, which generate an average output of more than nine watts during flight -- enough power to propel the miniature aircraft for one hour forty-seven minutes. Such aerial robots, equipped with information sensing and transmission capabilities, extend the sphere of awareness and mobility of human beings, and allow for surveillance or exploration of environments too hazardous or remote for human beings.

MAVs are expected to serve as an enabling technology for a plethora of civilian and military applications, including homeland security, reconnaissance, surveillance, tracking of terrorists/suspects, rescue and search, and highway/street patrol. With signal processing techniques (and other out-of-band techniques like visual perception which will not be discussed here), a team of three MAVs can locate the position of a target such as a person's or a car's communication interface. Due to the small size of MAVs, the tracking of MAVs is almost unnoticed by the target being tracked (Figure 6.1(b)). The velocity of an MAV is from 10 to 30 miles per hour, which is fast enough to track a human being or an automobile on local roads.

When a mobile ad hoc network is in operation, the mobile sensors carried by MAVs can eavesdrop routing messages and data traffic so to trace where a mobile wireless sender node is, infer the motion pattern of the mobile node, or identify a multi-hop path between a pair of nodes [51].

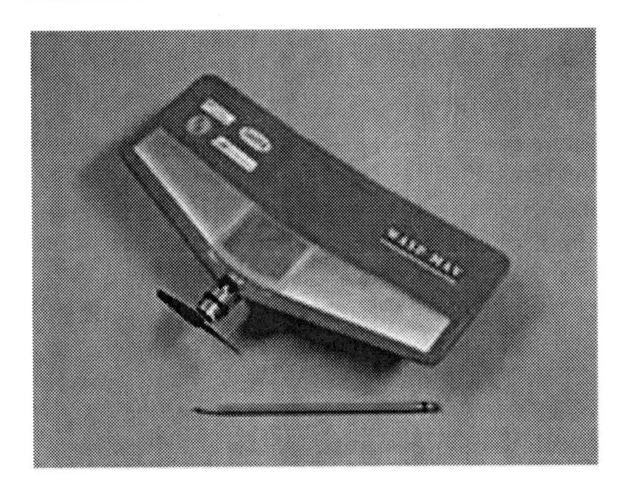

Fig. 6.1(a). Micro Aerial Vehicle (MAV)

Fig. 6.1(b). Street Patrol

6.1.2 On-demand routing

Most routing protocols in ad hoc networks fall into two categories: proactive routing and reactive routing (aka., on demand routing) [9]. In proactive ad hoc routing protocols like OLSR [1], TBRPF [34] and DSDV [53], mobile nodes constantly exchange routing messages which typically include node identities and their connection status to other nodes (e.g., link state or distance vector), so that every node maintains sufficient and fresh network topological information to allow them to find any intended recipients at any time. On the other hand, on demand routing has become a major trend in ad hoc networks. AODV [36, 37] and DSR [25] are common examples. Unlike their proactive counterparts, on demand routing operation is triggered by the communication demand at sources. Typically, an on demand routing protocol has two components: *route discovery* and *route maintenance*. In route discovery phase, the source establishes a route towards the destination by first flooding a route request (RREQ) message, and then receiving a route reply (RREP) sent by the destination. In the route maintenance phase, nodes on the route monitor the status of the forwarding path, and report to the source about route errors. Optimizations could lead to local repairs of broken links.

Clearly, transmitted routing messages and cached routing tables, if revealed to the adversary, will leak large amount of private information about the network. When this happens, proactive protocols and on-demand protocols show different levels of damages by design. With proactive routing, a compromised node has fresh topological knowledge about other proactive nodes during the entire network lifetime. It can also translate the topological map to a physical map using several anchor points (e.g., by techniques similar to sensor network's localization service [33, 46]). This way, a single-point of compromise allows the adversary to visualize the entire network and know where each node is. On the other hand, with on demand routing, the adversary has reduced chance in tracing the mobile network in the sense that only active routing entries are in cache and in transmission, and the traffic pattern is probabilistic (depending on application needs) and expires after a predefined timeout.

6.1.3 Overview

Recently, several on-demand anonymous routing schemes have been proposed to prevent mobile nodes from being traced by mobile sensors, including ANODR [27, 28, 50], AnonDSR [54], MASK [49], SDAR [8] and DisctANODR [52]. In this chapter, we illustrate the overhead incurred by security and anonymity operations of two distinct on-demand anonymous routing schemes among them, namely ANODR and SDAR. We use the standard on-demand scheme AODV [37] in the comparison to show how much overhead is paid by each anonymous on-demand scheme. Our simulation study shows that various design choices in anonymous routing trade performance with security protection. So far no anonymous routing scheme is able to surpass other competing schemes in all ad hoc scenarios studied.

The rest of the chapter is organized as follows. In Section 2 we review ANODR and SDAR protocols in detail. In Section 3 we evaluate their routing performance. Section 4 describes related work in wireless networks. Finally Section 5 summarizes the chapter.

6.2 Anonymous Routing Revisited

In this section we briefly overview anonymous routing approaches that do not use an on-demand design style, including the schemes that use global network knowledge and proactive routing approach in MANET. Then we review the two on-demand anonymous routing schemes that will be evaluated. We show the idiosyncrasies of each scheme and how the design choices affect routing protocol performance.

6.2.1 Anonymous routing *not* based on the on-demand approach

In wired network, anonymous routing design uses global knowledge about the network. Here we name it global-knowledge-based routing approach. In MANET, proactive routing protocols could also use the global-knowledge-based approach. Both are discussed below.

In global-knowledge-based routing approach, the network topology is fixed and pre-stored on each node. This includes the following designs. (1) In Chaum's DC-net [12], the network topology is suggested as a fixed and closed ring. (2) In Chaum's MIX-net [11], each message sender pre-stores the entire network topology, and then selects a random path from the known network topology in message routing. All subsequent MIX-net designs [6, 23, 26, 39] inherit this assumption. (3) In Crowds [43] and sorting network [41], all nodes are one logical hop away, pairwise communications exist with uniform cost. Anonymous messages are forwarded to the next node which is selected in a random manner. If this node is unavailable due to mobility or system crash, then another selection must be made following the same probabilistic method. In other words, every Crowds node or sorting network node is a member of an overlay network. Although at the network IP layer every node-to-node route is comprised of multiple IP routers, at the overlay layer such a node-to-node route is a single-hop logical link. This overlay anonymous network assumes either a global routing design or a proactive routing design at the IP network layer.

Nevertheless, static and global topology knowledge is no longer available in mobile ad hoc networks where the network topology constantly changes due to mobility, frequent route outage, and node joining/leaving. Maintaining the same global topology knowledge that is identical to fixed networks is very expensive and reveals the changing topological knowledge to node intruders.

In proactive routing approach, every node proactively and periodically exchanges routing messages with other nodes. Similar to the global routing approach, every node maintains fresh topology knowledge by paying routing communication overheads. In mobile ad hoc networks, various optimized proactive routing schemes, such as OLSR and TBRPF, have been proposed to reduce the incurred routing communication overheads. However, like their wired counterparts, the proactive ad-hoc routing schemes let every message sender maintain fresh topology knowledge about the network (even though the incurred communication overhead is less than their wired counterparts). Based on the proactively collected fresh routing knowledge, it is then possible to route anonymous messages to the next stop, which in turn routes the messages toward the final destination. This includes the following designs. (1) All MIX-nets can use proactive routing protocols at the network IP layer to acquire network topology knowledge, which is then used at the overlay MIX layer to route messages. (2) Like MIX-nets, an overlay of Crowds [43] or sorting network [41] can leverage proactive routing information as well. (3) In wired Internet, PipeNet [13] and Onion Routing [42] employ *anonymous virtual circuit* in routing and data forwarding. Every node knows its immediate previous stop (upstream node) and immediate next stop (downstream node). After a *signaling* procedure, a sequence of routing tables are created on the forwarding nodes to deliver data packets. Each routing table holds two columns of virtual circuit identifiers (VCI) in the form of "$vci_x \leftrightarrow vci_y$" [3]. If a node receives a packet with a vci_x presented in its routing table, the node then accepts the packet, overrides the stamp with the corresponding vci_y, and sends the changed packet to next stop (the source and the destination are denoted with special VCI tags). (4) In MIX route [24], a backbone network is formed to cover a mobile network. Every backbone node is a MIX, which uses proactive routing protocols to maintain fresh network topology of the backbone MIX-net.

In a nutshell, these global-knowledge-based routing and proactive routing schemes treat the underlying network as either a stationary graph, or fresh snapshots that can be treated as stationary graphs per proactive period. A shortcoming of applying these approaches in mobile networks comes from node intrusions. If adequate physical protection cannot be guaranteed for every mobile node, intrusion is inevitable within a time window. The adversary can thus compromise one mobile node, gather fresh network topology from the node's knowledge, then use simple localization schemes [33] to locate nearly all mobile nodes in the network.

Therefore, although various anonymous mechanisms, such as anonymous virtual circuit, MIX-net onion and backbone-style MIX-net remain effective in ad hoc networks, the global routing topology caching and proactive routing topology acquisition approaches are gradually replaced by the *on-demand* routing approach. Next we describe the recently-proposed on-demand anonymous routing schemes in the order of publication.

6.2.2 ANODR

ANODR [27, 28] is the first on-demand anonymous routing. Like PipeNet and Onion Routing, ANODR uses anonymous virtual circuit in routing and data forwarding. But unlike PipeNet and Onion Routing, every ANODR node does *not* know its immediate upstream node and immediate downstream node. Instead, ANODR is identity-free [50]. Each node only knows the physical presence of neighboring ad hoc nodes. This is achieved by a special anonymous signaling procedure.

Route discovery The source node initiates the anonymous signaling procedure. It creates an anonymous *global trapdoor* and an *onion* in a one-time route request (RREQ) flood packet.

1. *Anonymous global trapdoor*: The global trapdoor is a (semantically secure [17]) encryption of a well-known tag message (e.g., a predetermined bit-string "You are the destination") that can only be decrypted by the destination. Once the destination receives the flooded RREQ packet, it decrypts the global trapdoor and sees the well-known tag. But all other nodes see random bits after decryption. The design of global trapdoor requires anonymous end-to-end key agreement between the source and the destination.
2. *Onion*: As the RREQ packet is flooded from the source to the destination, each RREQ forwarding node adds a self-aware layer to the onion. Eventually the destination receives an onion that can be used to deliver a route reply (RREP) unicast packet back to the source.

The signaling procedure ends when the source receives RREP, and the anonymous virtual circuit is established during the RREP phase.

RREQ flood is a very expensive procedure, while public key crypto-processing is also expensive. According to measurement reports [10] on low-end mobile devices, common public key cryptosystems require 30–100 milliseconds of computation per encryption or per signature verification, 80--900 milliseconds of computation per decryption or per signature generation. Therefore, combining public key crypto and RREQ flood likely degrades routing protocol's performance. ANODR [27] proposes to avoid public key crypto except in the first RREQ flood between a pair of communicators.

In ANODR, each node is capable of doing encryption and decryption in both symmetric and public key cryptosystems. To establish the symmetric key shared between the source and the destination, the source must cache the *certified public key* of any intended destination prior to communication. (1) This implies that every network node must acquire a signed credential from an offline authority Ψ prior to network operations. The credential can be verified by the well-known PK_{Ψ}. The credential is in the form of "$[id, pk_{id}, validtime]_{SK\Psi}$" signed by SK_{Ψ}, where a unique network address *id* is assigned to a node, pk_{id} is the certified public key of the *id*, and *validtime* limits the valid period of the credential. Instead of using the unprotected plain *id*, the source remembers the credential and avoids using *id* in communication. (2) The credentials are not secret messages. They can be freely exchanged in the network to facilitate source nodes' caching experience. In contrast, the selection of a destination's pk_{id} is a secret random choice of the source node. (3) The selected pk_{id} of the destination is the global trapdoor key used in the *first* RREQ flood between the source and the destination. For better performance, a symmetric key is piggybacked in the first global trapdoor. Then the source would use the symmetric key in later global trapdoors between the same pair of source and destination. This spares the need of public key decryption in later RREQ floods.

At route reply (RREP) phase, the onion is decrypted to establish routing tables en route. When the onion comes back from the destination in the reverse order of encryption, the RREP upstream node chooses a random number *vci* and places it with the onion. The RREP downstream node receives this *vci,* then functions as the successive upstream node to choose its own *vci* and overrides the same field in the packet. As the RREP packet is processed and forwarded towards the source node, each route table on a forwarder *Y* records the VCIs in the form of "$vci_x \leftrightarrow vci_y$", where vci_x is chosen by *Y*'s RREP upstream node *X*, and vci_y is chosen by *Y* itself.

Data delivery ANODR seeks to make every data packet computationally one-time. This prevents traffic analysis and replay attacks. Hence a *vci* must be a *secret* shared on a forwarding hop. It is used as the cipher key to encrypt link frame payload (i.e., IP header and payload). Besides, the explicit VCIs stamped on data packets are computationally one-time. They are cryptographically strong pseudorandom sequences generated from the shared *vci,* which is now used as the shared secret *seed.* To share the secret *vci* on a hop, a per-hop key exchange scheme is needed. (1) At RREQ phase, an RREQ upstream node (which is later the RREP downstream) must put a one-time temporary public key in the RREQ flood packet. This one-time temporary public key is recorded by the RREQ downstream node (which is later the RREP upstream) for the source/destination session. The RREQ downstream node then overrides the field with its own temporary public key. (2) At RREP phase, the RREP upstream node (earlier the RREQ downstream) uses the stored one-time public key to encrypt the contents of RREP packet including the *vci* and the coming-back onion. If a one-hop RREP receiver decrypts the encrypted contents and sees the onion it sent out previously at RREQ phase, then this receiver (earlier the RREQ upstream) is en route. The anonymous virtual circuit is established when the source node receives the onion core it sent out a while ago. This way, the one-time public keys are plain data bits during RREQ floods. Per-hop key agreement overhead (using public key encryption/decryption) is paid during RREP unicasts.

Performance impact ANODR has to pay expensive public key crypto-processing overhead during the initial RREQ flood between a pair of communicators and all RREP unicasts. This significantly affects their routing performance. A variant of ANODR thus is to employ efficient Key Pre-distribution Schemes (KPS) to reach pairwise key agreement between two consecutive RREP forwarders. In a KPS scheme, the network needs an offline authority to initialize every node by loading appropriate personal key materials. Afterward, any two nodes can exchange key agreement material and agree on a key. If the underlying KPS scheme is a probabilistic one [15, 16] rather than a deterministic one [7], then the key agreement succeeds with high probability.

In addition, all the anonymous routing schemes reviewed in this section, i.e., ANODR and SDAR, have not implemented route optimization techniques specified in AODV and DSR (e.g., gratuitous route reply, proactive route fix using constrained flooding, etc.).

6.2.3 SDAR

SDAR [8] is a combination of proactive and on-demand route discovery. Unlike the purely on-demand ANODR, every SDAR node uses a *proactive* and *explicit* neighbor detection protocol to constantly see the snapshot of its one-hop mobile neighborhood. Every SDAR node periodically sends out a HELLO message holding the certified public key of the node. The SDAR HELLO messages are significantly longer than regular beacon messages because it holds long public keys (typically ≥1024-bit in a common public key cryptosystem like RSA and El Gamal).

An SDAR node is named as the *central node* as it sits at the center of its own one-hop transmission circle. A central node *X* explicitly sees its neighbors' network IDs and verifies associated credentials. *X* classifies its neighbors into three *trust levels* according to their behavior. Routing preference is given to the higher level nodes. This is implemented by group key management. *X* randomly chooses a key for all neighbors in the same trust level (except the lowest level, which is not protected by crypto-schemes). The key is then shared by *X* and these nodes. Routing messages intended for the highest level is encrypted with the group key corresponding to the highest level. Routing messages intended for the medium level is encrypted with either the group key corresponding to the medium level or the one corresponding to the highest level. Routing messages intended for the lowest level is not encrypted and thus seen by all listening nodes.

Route discovery SDAR also employs an on-demand route discovery procedure to establish ad hoc routes. Similar to ANODR, an SDAR source node *S* puts a global trapdoor in its RREQ flood packet. While the global trapdoor is encrypted with the destination *D*'s certified public key, a symmetric key is piggybacked into the global trapdoor to fulfill end-to-end key agreement. Nevertheless, unlike ANODR which uses ID-free tags, SDAR uses the destination *D*'s ID in the global trapdoor. This differentiates ANODR's ID-free global trapdoor from SDAR's ID-based global trapdoor.

Unlike ANODR, SDAR's RREQ flooding phase does not form any onion. Instead, the source node *S* puts its one-time public key *TPK* in the RREQ flood packet. *S* also piggybacks the corresponding one-time private key *TSK* in the global trapdoor, so that both *S* and *D* can decrypt any data encrypted by *TPK*. Each RREQ forwarder records *TPK*, chooses a random symmetric key *K*, and uses *TPK* to encrypt this per-stop *K*. This encrypted block is appended to the current RREQ packet. Finally when a RREQ packet reaches the destination *D* after traversing *l* hops, it contains *l* such appended *TPK*- encrypted blocks. *D* opens the global trapdoor and knows

TSK, then uses *TSK* to decrypt every *TPK*-encrypted block and thus shares a symmetric key with every forwarder of the received RREQ packet.

Similar to MIX-net, now the SDAR destination *D* has the *l* (symmetric) keys to form an RREP packet in the form of MIX-net onion. The destination *D* puts all symmetric key *K*s in the innermost core so that only the source *S* can decrypt the onion core and share *D*'s symmetric key with every RREP forwarder.

Once the source *S* receives the coming-back RREP, both the source *S* and the destination *D* have made a symmetric key agreement with every intermediate forwarder. Like the way RREP packet is delivered, *S* and *D* use MIX-net onion to deliver data payload to each other.

Data delivery The SDAR data delivery design uses layered encryption approach, which is similar to MIX-net's onion scheme.

Performance impact Compared to the purely on-demand ANODR, SDAR incurs extra neighbor detection overhead. Each neighbor detection message is significantly longer than short beacon messages, and also incurs a number of public key authentication and key exchange operations in the changing mobile neighborhood.

In on-demand route discovery, SDAR incurs excessive crypto-processing and communication overheads. Every RREQ forwarding must pay the cost of a public key encryption using *TPK*. This incurs expensive public key encryption overhead in the entire network per RREQ flood. SDAR's RREQ and RREP packets are very long. Each RREQ packet holds *l'* *TPK*-encrypted blocks where *l'* is the hop count from the source *S* to the current RREQ forwarder, each of the blocks is as long as the public key length. Every RREP packet and DATA packet has *l* MIX-net onion layers, each of the layers is at least 128-bit long (a typical symmetric key length).

6.2.4 Summary

Table 6.1 compares several design choices that may have significant impact on routing protocol performance and on security/performance tradeoffs.

Table 6.1. Protocol comparison

	ANODR	SDAR
Fully on-demand	Fully	Proactive nbr detection
PKC in RREQ flooding	First contact	All the time
Data delivery	Virtual circuit	Layered encryption
Neighbor exposure	No	Exposed
Dest. anonymity	Yes	Exposed

We compare the above aspects due to the following reasons. (1) Proactive neighbor detection incurs periodic communication and computational overheads on every mobile node. (2) Using expensive public key cryptography (PKC encryption/decryption) with expensive RREQ flood incurs intensive communication and computational overheads per flood. (3) In terms of data delivery performance, virtual circuit based schemes are more efficient than MIX-net's onion (layered encryption) based schemes. The latter one incurs l real-time encryption delay on the source node and then a single real-time decryption delay on every packet receiving nodes. (4) In SDAR, one-hop neighborhood is exposed to internal (and possibly external) adversary. This is not a security problem in fixed networks. But in mobile networks, this reveals the changing local network topology to mobile traffic sensors, which can quickly scan the entire network for once and obtain an estimation of the entire network topology. (5) Recipient anonymity (of the destination's network ID) is a critical security concern. Otherwise, every RREQ receiver can see how busy a destination node is. This traffic analysis can be used by the mobile traffic sensors to define the priority in node tracing.

6.3 Performance Evaluation

The performance of the anonymous ad-hoc routing protocols discussed here is evaluated through simulation.

In the evaluation, the aforementioned anonymous ad-hoc routing protocols are presented for comparison together with the original AODV. Our evaluation concerns the influence from processing overhead incurred by the cryptosystems in use and also the influence of routing control overhead caused by different size of routing control packets. The simulation of the protocols are all implemented based on AODV. Each of them implements the main principles but uses different cryptosystems in establishing the secret hop key.

The cryptosystems include the public key cryptography and a variant of efficient Key Pre-distribution Schemes (KPS). In a public key scheme, the network needs an offline authority to grant every network member a credential signed by the authority's signing key, so that any node can verify a presented credential with the authority's well-known public key. The standard ANODR and SDAR use public key cryptography. In a KPS scheme, the network needs an offline authority to load every node with personal key materials. Afterward, any two nodes can use their key materials and agree on a symmetric key. A variant of ANODR using KPS (in RREP unicasts) is tested in our simulation study. It uses the probabilistic KPS scheme [15] (denoted as ANODR-KPS). In ANODR-KPS, the probability of achieving a successful key agreement at each hop is 98%. In other words, key *vci* agreement fails with 2% at every RREP hop. A new route discovery procedure will be invoked eventually by the source.

We evaluate the performance of these protocols in terms of the following metrics. (i) *Packet delivery fraction* -- the ratio between the number of data packets received and those originated by the sources. (ii) *Average end-to-end data packet latency* -- the time from when the source generates the data packet to when the destination receives it. This includes: route acquisition latency, processing delays at various layers of each node, queueing at the interface queue, retransmission delays at the MAC, propagation and transfer times. (iii) *Average route acquisition latency* -- the average latency for discovering a route, i.e., the time elapsed between the first transmission of a route request and the first reception of the corresponding reply. (iv) *Normalized control packet overhead* -- the number of routing control **packets** transmitted by a node normalized by number of delivered data packets, averaging over all the nodes. Each hop-wise transmission of a routing packet is counted as one transmission. (v) Normalized control byte overhead -- the total **bytes** of routing control packets transmitted by a node normalized by delivered data bytes, averaging over all the nodes. Each hop-wise transmission of a routing packet is counted as one transmission. This metric is useful in evaluating the extra control overhead of ANODR-KPS. With these metrics, the overall network performance is observed by *packet delivery fraction.* Influence from processing delay and packet size can be validated through latency metrics and overhead metrics. In addition, SDAR requires each node to periodical broadcast messages to neighboring one-hop nodes. When we compare the five performance metrics, we leave out the periodical routing control overhead for SDAR and study it in a separate discussion.

6.3.1 Crypto-processing performance measurement

The processing overhead used in our simulation is based on actual measurement on low-end devices. Table 6.2 shows our measurements on the performance of different cryptosystems. For public key cryptosystems, the table shows processing latency per operation. For symmetric key cryptosystems (the five AES final candidates), the table shows encryption/decryption bit-rate.

Table 6.2. Processing overhead of various cryptosystems (on iPAQ3670 pocket PC with Intel StrongARM 206MHz CPU)

Cryptosystem	decryption	encryption
ECAES (160-bit key)	42ms	160ms
RSA (1024-bit key)	900ms	30ms
El Gamal (1024-bit key)	80ms	100ms
AES/Rijndael (128-bit key & block)	29.2Mbps	29.1Mbps
RC6 (128-bit key & block)	53.8Mbps	49.2Mbps
Mars (128-bit key & block)	36.8Mbps	36.8Mbps
Serpent (128-bit key & block)	15.2Mbps	17.2Mbps
TwoFish (128-bit key & block)	30.9Mbps	30.8Mbps

Clearly, different cryptosystems introduce different processing overhead, thus have different impact on anonymous routing performance. For all public key cryptographic operations in the simulation, we use ECAES with 160-bit key. For the symmetric cryptography, we use AES/Rijndael with 128-bit key and block. The coding bandwidth is about 29.2Mbps. As an example, in ANODR, computational delay is approximately 0.02ms for each onion construction during each RREQ and RREP forwarding, and another public key processing time 160+42=202ms for RREP packets. The KPS based ANODR trades link overhead for processing time, i.e., ANODR-KPS uses 1344 bits and 1288 bits key agreement material for RREQ and RREP packets respectively. Each of them requires only 1ms extra time in processing packets.

6.3.2 Simulation model

The simulation is performed in QualNet™[45], a packet level simulator for wireless and wired networks developed by Scalable Network Technologies Inc. The distributed coordination function (DCF) of IEEE 802.11 is used as the MAC layer in our experiments. It uses Request-To-Send (RTS) and Clear-To-Send (CTS) control packets to provide virtual carrier sensing for unicast data packets to overcome the well-known hidden terminal problem. Each unicast data transmission is followed by an ACK. The radio uses the *two-ray ground reflection* propagation model and has characteristics similar to commercial radio interfaces (e.g., WaveLAN). The channel capacity is 2Mbps.

The network field is 2400m×600m with 150 nodes initially uniformly distributed. The transmission range is 250m. *Random Way Point* (RWP) model is used to simulate node mobility. In our simulation, the mobility is controlled in such a way that minimum and maximum speeds are always the same (to fix the speed decay problem [48]), but increase from 0 to 10 m/sec in different runs. The pause time is fixed to 30 seconds. CBR sessions are used to generate network data traffic. For each session, data packets of 512 bytes are generated in a rate of 4 packets per second. The source-destination pairs are chosen randomly from all the nodes. During 15 minutes simulation time, a constant, continuously renewed load of 5 short-lived pairs is maintained. All simulations are conducted in identical network scenarios (mobility, communication traffic) and routing configurations across all schemes in comparison. All results are averaged over multiple runs with different seeds for the random number generator.

6.3.3 Routing performance measurement

Here we present the results and our observations. Figure 6.2 shows the comparison of packet delivery ratio. In an environment without any attackers, the original AODV protocol indicates the best performance possible on this metric. ANODR-KPS has the similar performance with the original AODV, as it only uses efficient symmetric cryptography when exchanging routing packets, effectively accelerating the route discovery process and making the established routes more durable. ANODR results in degradation in delivery ratio, primarily caused by the longer delay required for asymmetric key encryption/decryption in the route reply phase. SDAR shows quicker decreasing trend due to the reasons that it requires asymmetric key encryption/decryption in both route request and route reply phases, and hop-related public key encryption/decryption at the destination nodes. In a mobile environment, excessive delay in route discovery process makes it harder to establish and maintain routes. All of the curves show a more or less yet steady descendant when mobility increases. This is natural as increasing mobility will cause more packet loss.

Figure 6.3 illustrates the data packet latency. As SDAR uses public key cryptography throughout the round trip of route discovery, a node needs to wait longer time before a route is established. ANODR shows a shorter average data packet latency because it only uses public key encryption/decryption when forwarding route reply messages. ANODR-KPS has nearly the same data packet delay with the original AODV, thanks to the efficient symmetric encryption algorithms and hash functions used. When there is little mobility, all protocols display small data packet latency, because once a route is established, a stable network allows a longer average route lifetime. When mobility increases, data packet latency increases accordingly. The latency trend slows down in high mobility because more data packets are lost due to mobility. Packets with shorter routes are more likely to survive and be delivered successfully.

Figure 6.4 shows the average route acquisition delay under different node mobility. It validates that the latency for establishing a route using SDAR is longer than that of other protocols.

Figure 6.5 compares the number of normalized control packets over all of the protocols. All of the anonymous ad-hoc protocols have higher numbers of normalized control packets than that of the original AODV due to the fact of the less delivered data packets and the route re-discovery because of route errors. Figure 6.6 compares the normalized control overhead in terms of bytes. It's clear that ANODR-KPS incurs much more overhead. This is expected as the size of the control packets (RREQ and RREP, primarily) of ANODR-KPS is about two times or more as that of ANODR or SDAR, and three times or more as that of the original AODV.

Figure 6.7 reports the overhead of the proactive key establishment of SDAR. It shows the normalized *number* and *bytes* of neighbor authentication packets under different mobility condition. SDAR uses periodical hello messages containing public keys for community management. Thus the number of periodical control packets are not affected by mobility. However, since the number of packets delivered decreases as the mobility increases, the overhead packets increases gradually when mobility increases (the scale is given at the left side of Figure 6.7). Similar trend for overhead measured in bytes is observed (the scale is shown at the right side of Figure 6.7). On the other hand, the number of authentication packets are determined by the frequency of the Hello message. In this simulation we use the default AODV Hello frequency, i.e., one Hello message per second. Compared with the normalized routing overhead presented in Figures 6.5 and 6.6, the current periodic packet overhead is at a similar level as the overhead generated by the route discovery and maintenance (Figure 6.5). Reduction of these neighbor authentication overhead could be achieved through possible adaption on Hello interval. However, SDAR has a lower level of normalized authentication bytes than its routing control bytes (Figure 6.6). This is because that the size of Hello message is smaller than the sizes of RREQ and RREP packets in SDAR.

In summary, our main findings are: (i) Control packet size, if controlled within a reasonable size, has less impact on performance. E.g., Figure 6.2

shows close delivery ratios of AODV and ANODR-KPS. But ANODR-KPS has much higher control bytes as shown in Figure 6.6. (ii) Processing delay has great impact on delivery ratio in a mobile environment. E.g., ANODR-KPS and SDAR have close combined packet size, but Figure 6.2 shows that their delivery ratios have large difference.

On the other hand, the simulation results demonstrate the existence of trade-offs between routing performance and security protection. Because the ad hoc route discovery (RREQ/RREP) procedure is *time critical* in a mobile network, excessive crypto-processing latency would result in stale routes and hence devastate routing performance. Our results show while ANODR could be suitable for low-end nodes and medium mobility, SDAR are better when used with high-end nodes that can run public key cryptography efficiently. In order to design a practical anonymous ad hoc routing scheme, we must find out the optimal balance point that can both avoid expensive cryptographic processing and provide needed security protection at the same time.

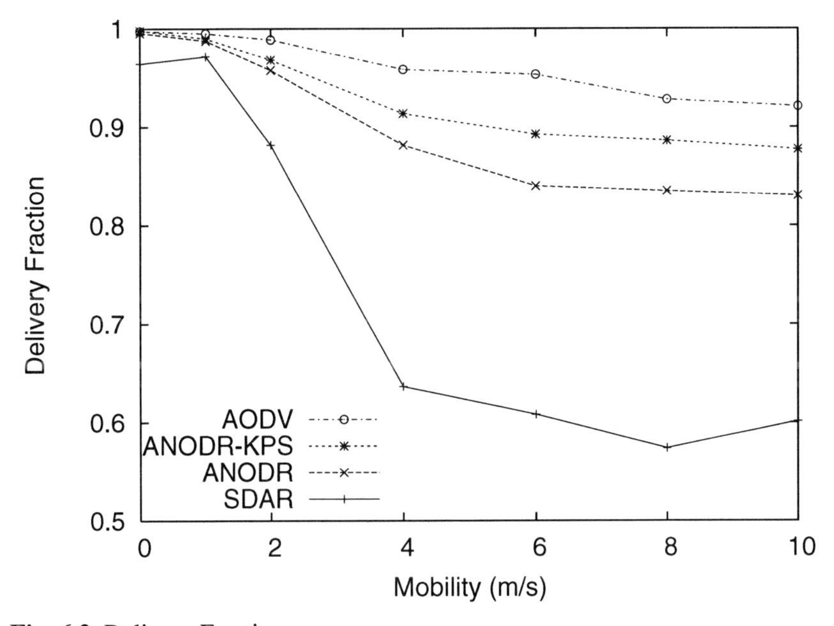

Fig. 6.2. Delivery Fraction

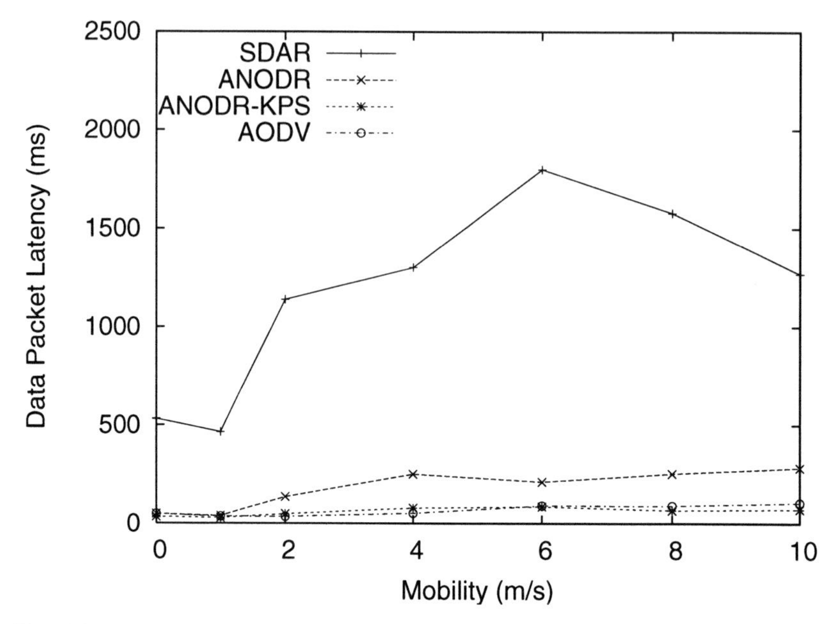

Fig. 6.3. Data Packet Latency (ms)

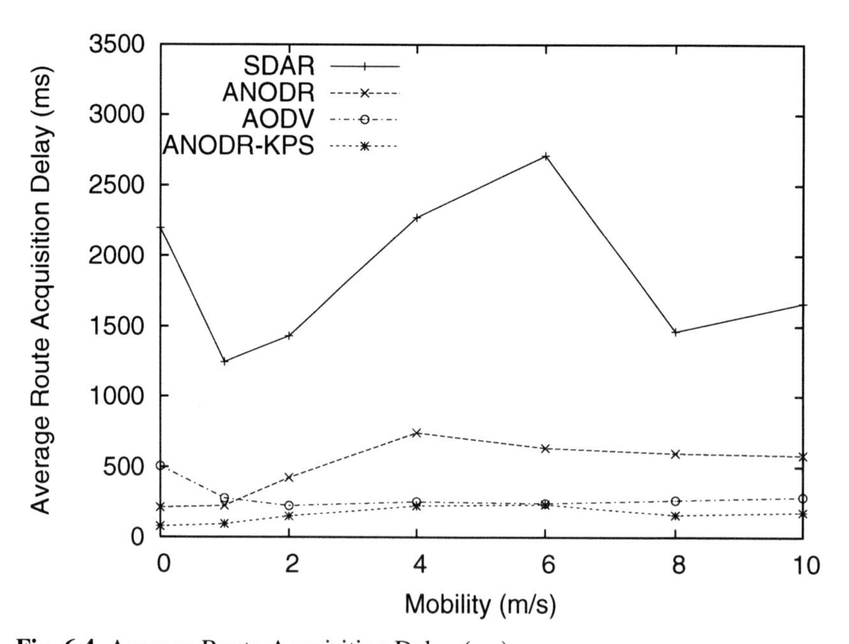

Fig. 6.4. Average Route Acquisition Delay (ms)

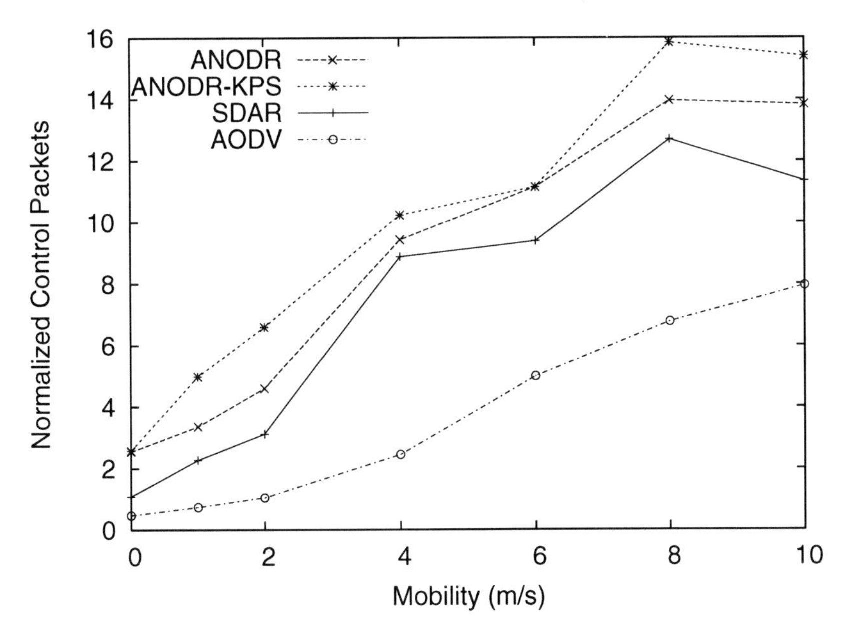

Fig. 6.5. Normalized Control Packets

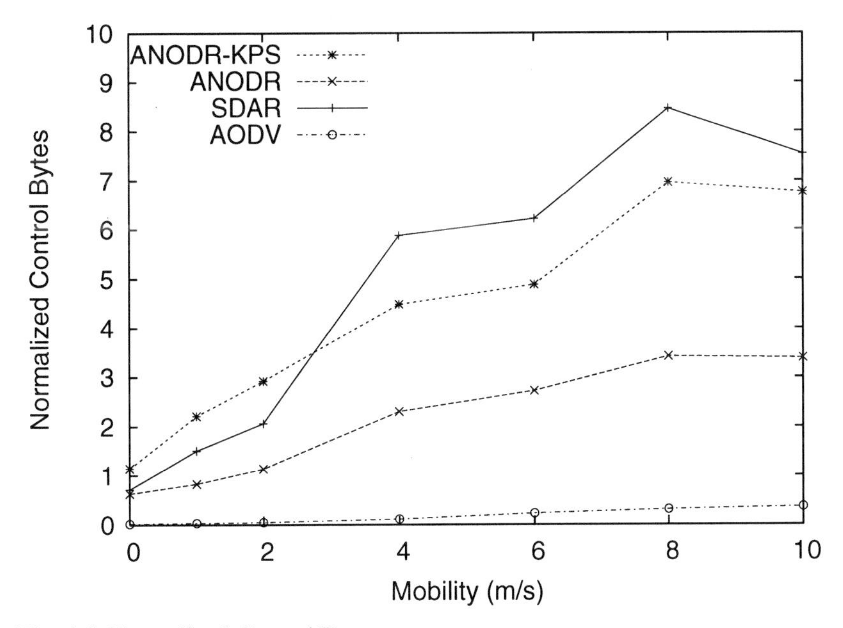

Fig. 6.6. Normalized Control Bytes

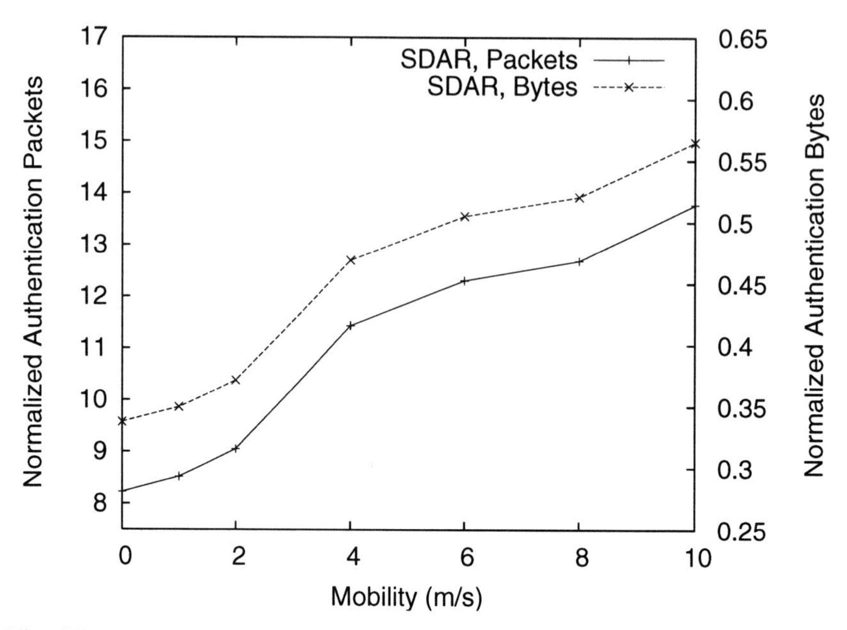

Fig. 6.7. SDAR Normalized Neighbor Authentication Overhead

6.4 Related Work

Existing anonymity protection schemes for wireless networks fall into a spectrum of classes. In "last hop" wireless networks (including cellular networks and wireless LANs), the demand of user roaming requires more promising assurance on the privacy of mobile users. The network participants considered in related research are typically the mobile users, the home servers of the users, the foreign agent servers local to the users, and the eavesdroppers (could be other mobile users). In [2, 44], mobile users are associated with dynamic *aliases* that appear unintelligible to anyone except the home server. Then the foreign agent server accepts the user's connections upon the home server's request. In [19], mobile users employ Chaum's *blind signature* to establish authenticated but anonymous connections to the foreign agent server. Hu and Wang [20] propose to use *anonymous rendezvous*, an anonymous bulletin board, to let mobile nodes anonymously connect to their communicators. These efforts provide unlinkability protections between node identities and their credentials during anonymous transactions. This design goal is orthogonal to anonymous on-demand routing.

In wireless sensor networks, distributed sensor nodes monitor target events, function as information sources and send sensing reports to a number of sinks (command center) over multi-hop wireless paths. The sensor nodes and sinks are typically stationary in WSN. Deng et al. [14] propose to use multi-path routes and varying traffic rates to protect recipient anonymity of the network sinks. Ozturk et al. [35] propose to prevent a mobile adversary (e.g., a poacher) from tracing a sensor report packet flow back to a mobile target's location (e.g., a panda). The sensor nodes must report the mobile target's status to the sinks via *phantom flooding*, which is a sequential combination of random walk and controlled flooding. Both proposals seek to prevent the adversary from tracing network packet flows back to the sources or the sinks. In these proposals, routers (i.e., forwarding nodes) are stationary. They are not applicable to a network in which every router is mobile.

In geographic services, both Location-Base Services [18] and Mix Zones [5] study how to use middleware services to ensure location privacy with respect to time accuracy and position accuracy. They study user anonymity protection in static "geographic regions" with boundary lines. The regions are fixed during the network lifetime, and anonymity protection degrades in a single region. Besides, since the anonymity protection stops at the middleware layer (typically above the network IP layer), the adversary can trace a mobile node using network identities/addresses at the network layer and the link layer, or radio signatures at the physical layer. The protection of upper layer user identities by these middleware is not the focus of anonymous routing.

6.5 Conclusion

In this chapter we have illustrated the connections of two types of recently-proposed on-demand anonymous routing schemes through two examples, namely ANODR and SDAR. We analyze various factors that affect their routing performance and security. We further demonstrate that tradeoffs exist between the performance and the degree of protection. Our simulation study verifies that various choices in anonymous routing design have significant impact on anonymous routing protocol performance. Our results show that ANODR is suitable in mobile ad hoc networks with heterogeneous nodes (including low-end nodes) and medium mobility. SDAR is suitable in mobile ad hoc networks with high-end nodes that can run public key cryptographic operations efficiently. We conclude that more extensive performance study is needed to evaluate the practicality of the proposed anonymous proposals, the enhancements of them, and the new anonymous routing schemes.

References

1. C. Adjih, T. Clausen, P. Jacquet, A. Laouiti, P. Minet, P. Muhlethaler, A. Qayyum, and L. Viennot. Optimized Link State Routing Protocol. Internet Draft.
2. G. Ateniese, A. Herzberg, H. Krawczyk, and G. Tsudik. Untraceable Mobility or How to Travel *Incognito*. *Computer Networks*, 31(8):871–884, 1999.
3. ATM Forum. Asynchronous Transfer Mode. http://www.atmforum.org/.
4. D. Balfanz, G. Durfee, N. Shankar, D. K. Smetters, J. Staddon, and H.-C. Wong. Secret Handshakes from Pairing-Based Key Agreements. In *IEEE Symposium on Security and Privacy*, pages 180–196, 2003.
5. A. R. Beresford and F. Stajano. Location Privacy in Pervasive Computing. *IEEE Pervasive Computing*, 2(1):46–55, 2003.
6. O. Berthold, H. Federrath, and S. K¨opsell. Web MIXes: A system for anonymous and unobservable Internet access. In H. Federrath, editor, *DIAU'00, Lecture Notes in Computer Science 2009*, pages 115–129, 2000.
7. R. Blom. An Optimal Class of Symmetric Key Generation System. In T. Beth, N. Cot, and I. Ingemarsson, editors, *EUROCRYPT' 84, Lecture Notes in Computer Science 209*, pages 335–338, 1985.
8. A. Boukerche, K. El-Khatib, L. Xu, and L. Korba. SDAR: A Secure Distributed Anonymous Routing Protocol for Wireless and Mobile Ad Hoc Networks. In *29th IEEE International Conference on Local Computer Networks (LCN'04)*, pages 618–624, 2004.
9. J. Broch, D. A. Maltz, D. B. Johnson, Y.-C. Hu, and J. Jetcheva. A Performance Comparison of Multi-Hop Wireless Ad Hoc Network Routing Protocols. In *ACM MOBICOM*, pages 85–97, 1998.
10. M. Brown, D. Cheung, D. Hankerson, J. L. Hernandez, M. Kurkup, and A. Menezes. PGP in Constrained Wireless Devices. In *USENIX Security Symposium (Security '00)*, 2000.
11. D. L. Chaum. Untraceable electronic mail, return addresses, and digital pseudonyms. *Communications of the ACM*, 24(2):84–88, 1981.
12. D. L. Chaum. The Dining Cryptographers Problem: Unconditional Sender and Recipient Untraceability. *Journal of Cryptology*, 1(1):65–75, 1988.
13. W. Dai. PipeNet 1.1. http://www.eskimo.com/_weidai/pipenet.txt, 1996.
14. J. Deng, R. Han, and S. Mishra. Intrusion Tolerance and Anti-Traffic Analysis Strategies for Wireless Sensor Networks. In *IEEE International Conference on Dependable Systems and Networks (DSN)*, pages 594–603, 2004.
15. W. Du, J. Deng, Y. S. Han, and P. K. Varshney. A Pairwise Key Pre-distribution Scheme for Wireless Sensor Networks. In *ACM CCS*, pages 42–51, 2003.
16. L. Eschenauer and V. D. Gligor. A Key-Management Scheme for Distributed Sensor Networks. In *ACM CCS*, pages 41–47, 2002.
17. S. Goldwasser and S. Micali. Probabilistic Encryption. *Journal of Computer and System Sciences*, 28(2):270–299, 1984.
18. M. Gruteser and D. Grunwald. Anonymous Usage of Location-Based Services Through Spatial and Temporal Cloaking. In *MobiSys03*, 2003.
19. Q. He, D. Wu, and P. Khosla. Quest for Personal Control over Mobile Location Privacy. *IEEE Communications Magazine*, 42(5):130–136, 2004.

20. Y.-C. Hu and H. J. Wang. A Framework for Location Privacy in Wireless Networks. In *ACM SIGCOMM Asia Workshop*, 2005.
21. P. G. Ifju, S. M. Ettinger, D. Jenkins, Y. Lian, W. Shyy, and M. Waszak. Flexible-wing-based Micro Air Vehicles. In *40th AIAA Aerospace Sciences Meeting*, 2002.
22. P. G. Ifju, S. M. Ettinger, D. Jenkins, and L. Martinez. Composite materials for Micro Air Vehicles. *SAMPE Journal*, 37(4):7–13, 2001.
23. A. Jerichow, J. M¨uller, A. Pfitzmann, B. Pfitzmann, and M. Waidner. Real-Time MIXes: A Bandwidth-Efficient Anonymity Protocol. *IEEE Journal on Selected Areas in Communications*, 16(4), 1998.
24. S. Jiang, N. Vaidya, and W. Zhao. A MIX Route Algorithm for Mix-net in Wireless Ad hoc Networks. In *IEEE International Conference on Mobile Ad-hoc and Sensor Systems (MASS)*, 2004.
25. D. B. Johnson and D. A. Maltz. Dynamic Source Routing in Ad Hoc Wireless Networks. In T. Imielinski and H. Korth, editors, *Mobile Computing*, volume 353, pages 153–181. Kluwer Academic Publishers, 1996.
26. D. Kesdogan, J. Egner, and R. Buschkes. Stop-and-go MIXes Providing Probabilistic Security in an Open System. *Second International Workshop on Information Hiding (IH'98), Lecture Notes in Computer Science 1525*, pages 83–98, 1998.
27. J. Kong. *Anonymous and Untraceable Communications in Mobile Wireless Networks*. PhD thesis, University of California, Los Angeles, June 2004.
28. J. Kong and X. Hong. ANODR: ANonymous On Demand Routing with Untraceable Routes for Mobile Ad-hoc Networks. In *ACM MOBIHOC'03*, pages 291–302, 2003.
29. J. Kong, X. Hong, and M. Gerla. A New Set of Passive Routing Attacks in Mobile Ad Hoc Networks. In *IEEE MILCOM*, 2003.
30. F. J. MacWilliams and N. J. A. Sloane. *The Theory of Error-Correcting Codes*. Amsterdam, The Netherlands, North-Holland, 1988.
31. R. Motwani and P. Raghavan. *Randomized algorithms*. Cambridge University Press, 1995.
32. National Institute of Standards and Technology. Advanced Encryption Standard. http://csrc.nist.gov/encryption/aes/, 2001.
33. D. Niculescu and B. Nath. Ad hoc positioning system (APS). In *IEEE GLOBECOM*, 2001.
34. R. Ogier, M. Lewis, and F. Templin. Topology Dissemination Based on Reverse-Path Forwarding (TBRPF). http://www.ietf.org/internet-drafts/draft-ietf-manet-tbrpf-07.txt, March 2003.
35. C. Ozturk, Y. Zhang, and W. Trappe. Source-Location Privacy in Energy-Constrained Sensor Network Routing. In *ACM SASN*, pages 88–93, 2004.
36. C. E. Perkins and E. M. Royer. Ad-Hoc On-Demand Distance Vector Routing. In *IEEE WMCSA'99*, pages 90–100, 1999.
37. C. E. Perkins, E. M. Royer, and S. Das. Ad-hoc On Demand Distance Vector (AODV) Routing. http://www.ietf.org/rfc/rfc3561.txt, July 2003.
38. A. Pfitzmann and M. K¨ohntopp. Anonymity, Unobservability, and Pseudonymity - A Proposal for Terminology. In H. Federrath, editor, *DIAU'00, Lecture Notes in Computer Science 2009*, pages 1–9, 2000.

39. A. Pfitzmann, B. Pfitzmann, and M. Waidner. ISDNMixes:Untraceable Communication with Very Small Bandwidth Overhead. In *GI/ITG Conference: Communication in Distributed Systems*, pages 451–463, 1991.
40. A. Pfitzmann and M. Waidner. Networks Without User Observability: Design Options. In F. Pichler, editor, *EUROCRYPT' 85, Lecture Notes in Computer Science 219*, pages 245–253, 1986.
41. C. Rackoff and D. R. Simon. Cryptographic defense against traffic analysis. In *Symposium on the Theory of Computation (STOC)*, pages 672–681, 1993.
42. M. G. Reed, P. F. Syverson, and D. M. Goldschlag. Anonymous Connections and Onion Routing. *IEEE Journal on Selected Areas in Communications*, 16(4), 1998.
43. M. K. Reiter and A. D. Rubin. Crowds: Anonymity for Web Transactions. *ACM Transactions on Information and System Security*, 1(1):66–92, 1998.
44. D. Samfat, R. Molva, and N. Asokan. Untraceability in Mobile Networks. In *ACM MOBICOM*, pages 26–36, 1995.
45. Scalable Network Technologies (SNT). QualNet. http://www.qualnet.com/.
46. Y. Shang, W. Ruml, Y. Zhang, and M. P. J. Fromherz. Localization from Mere Connectivity. In *ACM MOBIHOC*, pages 201–212, 2003.
47. G. S. Vernam. Cipher Printing Telegraph Systems for Secret Wire and Radio Telegraphic Communications. *Journal American Institute of Electrical Engineers*, XLV:109–115, 1926.
48. J. Yoon, M. Liu, and B. Noble. Sound Mobility Models. In *ACM MOBICOM*, pages 205–216, 2003.
49. Y. Zhang, W. Liu, and W. Lou. Anonymous Communications in Mobile Ad Hoc Networks. In *IEEE INFOCOM*, 2005.
50. J. Kong, X. Hong, M. Gerla. An Identity-free and On Demand Routing Scheme against Anonymity Threats in Mobile Ad-hoc Networks. IEEE Transactions on Mobile Computing, to appear 2007.
51. X. Hong, J. Kong, M. Gerla, "Mobility Changes Anonymity: New Passive Threats in Mobile Ad Hoc Networks", Wireless Communications & Mobile Computing (WCMC), Special Issue of Wireless Network Security, Vol. 6, Issue 3, May 2006, Page(s):281–293.
52. L. Yang, M. Jakobsson, S. Wetzel. "Discount Anonymous On Demand Routing for Mobile Ad hoc Networks," in the proceedings of *SECURECOMM '06.*
53. C. E. Perkins and P. Bhagwat, Highly Dynamic Destination-Sequenced Distance-Vector Routing (DSDV) for Mobile Computers, ACM SIGCOMM, pp. 234–244, 1994.
54. R. Song and L. Korba and G. Yee. AnonDSR: Efficient Anonymous Dynamic Source Routing for Mobile Ad-Hoc Networks, ACM Workshop on Security of Ad Hoc and Sensor Networks (SASN), 2005.

7 Computer Ecology: Responding to Mobile Worms with Location-Based Quarantine Boundaries

Baik Hoh[1] and Marco Gruteser[2]

[1] WINLAB, ECE Department
Rutgers, The State University of New Jersey
[2] WINLAB, ECE Department
Rutgers, The State University of New Jersey

7.1 Introduction

A current trend in pervasive devices is towards multi-radio support, allowing direct local interaction between devices in addition to maintaining long-haul links to infrastructure networks. Many current cell phones already contain Bluetooth radios that enable peer-to-peer exchange of files and usage of services from nearby devices. Bluetooth is also available in some automobiles and the US Federal Communications Commission has reserved spectrum for Dedicated Short Range Communications (DSRC), a wireless communications standard for inter-vehicle networks based on the IEEE 802.11 medium access protocol [1]. Example applications are collaborative crash warning and avoidance, dynamic traffic light control, or ad hoc forwarding of traffic probe information [2, 3].

Unfortunately, peer-to-peer interaction between devices provides an alternative propagation path for worms and virus [4, 5]. The Internet experience illustrates that worm attacks are a significant concern and a proof-of-concept Bluetooth worm, Cabir, has already been implemented.[3] More aggressive worms that exploit bugs (e.g., buffer overflow in bluetooth software/protocol stack [7, 8]) and make unwanted phone calls are not hard to imagine [9, 10], and likely as financial incentives increase. More recently, several research articles [4, 11–13] warn that worms and viruses could cause denial-of-service or energy-depletion attacks.

Regardless of the sophistication of the prevention strategies, in an environment with high reliability requirements it is only prudent to also plan for outbreaks with appropriate containment strategies. Peer-to-peer replication over short-range wireless networks creates a challenge for intrusion detection and response, because the worm cannot be observed and blocked by intrusion detection and response systems

[3] In fact, a Cabir outbreak was recently reported during a sporting event at the Helsinki Olympic Stadium [6] and rumors are abound that it could spread to in-car computers of a luxury sport utility vehicles.

in the cellular service provider's core network. Instead intrusion detection must be deployed on resource-constrained mobile devices or on specialized honeypot devices distributed in high-traffic zones [14, 15]. Regardless of the employed intrusion detection method, these constraints will lead to a delay between the time of outbreak and alarm because of distributed processing delays and human analysis. Thus, the intrusion response system only has at best an outdated few of the current worm propagation.

In this work, we consider an intrusion response architecture where a service provider remotely administers mobile nodes over the wide-area infrastructure wireless network. Using ecologically inspired location-based quarantine boundary estimation techniques, the service provider can estimate a set of likely infected nodes. This allows the service provider to concentrate efforts on infected nodes and minimize inconvenience and danger to non-affected parties.

The remainder of this paper is structured as follows. Section 7.2 clarifies threat model and system assumptions. It also defines the estimation problem that this paper addresses. Section 7.3 develops a quarantine boundary estimation algorithm from ecological diffusion-reaction and advection models. We evaluate our proposed algorithm by applying it to two ad hoc network scenarios: a pedestrian random-walk and an a vehicular network on a highway. These results are reported in section 7.4. In section 7.5, we analyze the simulation results and discuss the effectiveness of the approach. In addition, we discuss how to locate *Patient 0* based on a set of intrusion reports. Section 7.6 compares our work with directly related prior works before we conclude.

7.2 Threat Assessment

We consider a network system that comprises mobile radio nodes with ad hoc networking capabilities and a wide-area wireless infrastructure network with central network management by a service provider. Each mobile node is connected to the infrastructure network, provided that radio coverage is available, and can directly communicate with other mobile nodes over a short-range radio interface. Examples of such a system are a CDMA/GSM cell-phone network with Bluetooth handsets or an automotive telematics system supporting CDMA and DSRC. We assume that the service provider can locate each mobile node. This could be implemented through Assisted GPS on the nodes or triangulation technology in the infrastructure. Hybrid approaches are also possible.

In this network system, worms and viruses may spread through ad hoc connections over the shortrange interface, rather than the infrastructure network. Mobile nodes can be infected if they are a neighbor, meaning in the communication range Cr, of an already infected node. Typically, an infected node is able to identify its neighbors through network discovery mechanisms (e.g., IEEE 802.11 probe request, probe response protocol) or by monitoring communications within its range. Not all neighbors must be susceptible to the attack because an attack might depend on a vulnerability in a particular implementation or the configuration of the device.[4] We can assume, however, that malware will infect these susceptible nodes through software vulnerabilities soon after they first enter the communication range of an infected node. On Bluetooth networks, the BlueSmack attack [16] already provides an example of malware that exploits a buffer overflow vulnerability in a Bluetooth implementation. BlueSmack sends an oversized L2CAP echo request packet to a Bluetooth host to overflow the allocated receive buffer. While this attack only crashes the Bluetooth stack, similar vulnerabilities will probably allow future malware to execute arbitrary code. Even though any specific realization of such an attack is to date unknown, this will likely allow malware to spread without any user intervention through software exploits, similar the spread of worms among Internet hosts.

Malware spreading over the ad hoc network is more difficult to detect and contain than malware spreading over an infrastructure network, because the network does not contain concentration points (choke-points) where centralized intrusion detection and traffic filtering techniques can be applied. Instead detection and response techniques must be implemented in a highly distributed architecture on the mobile nodes themselves. While it is plausible that malware propagates over both the short-range *and* the infrastructure network [5], we ignore this case here because the infrastructure connections can be prevented with traditional defenses.

We are especially concerned with unknown malware, which signature-based intrusion detection systems cannot yet detect. The service provider may learn a new epidemic through different mechanisms ranging from mundane user calls to its service hotline to a sophisticated anomaly detection system. We observe that any of these mechanisms suffer from a high false-alarm probability and thus require the intervention of human analyst to verify that an actual outbreak exists. This leads to a detection delay of minutes in the best case. Even in a fully automated system, a distributed intrusion detection system would add delay due to the distributed detection processing and the latency overhead of delay-tolerant communication. During this time the malware can spread further (and anomaly reports from new nodes may again require verification) leaving the analyst with an incorrect, delayed view of the epidemic.

This work assumes, however, that the analyst can accurately locate *patient 0*, the initially infected node. If every node runs an intrusion detection system with sufficient memory for logging events, the infection can generally be traced to its origin. An inaccurate estimate of *patient 0*'s position will lead to degraded system performance. We will discuss more about how to locate *patient 0* from multiple intrusion reports of intrusion detection systems in section 7.5. We leave making the system more robust to the *patient 0* estimate for future work.

[4] In particular, settings such as the Bluetooth non-discoverable mode might provide limited protection against some attacks while other brute force scan mechanisms are still possible [10].

In summary, the service provider will determine from a range of clues whether an intrusion took place. The service provider characterizes an intrusion by a tuple $(pos_x, pos_y, time)$ that describes the time and position of *patient 0* at the start of the outbreak.

7.2.1 Intrusion response

Given that an intrusion event occurred, a service provider's main interest lies in minimizing inconvenience and potential danger (e.g., users may depend on cell phones for 911/112 emergency calls or distractions from an infected in-vehicle system may cause car accidents) to customers.

Responding effectively requires a secure management interface to the mobile nodes that allows service providers to remotely regain control of a compromised mobile node. Remote management interfaces are common practice for managing servers in larger data centers and have become increasingly prevalent in the cell phone world. For example, the Open Mobile Alliance Client Provisioning Architecture [17] allows over-the-air configuration of mobile nodes. It also specifies a privileged configuration context, whose settings cannot be modified by users or applications. Such interfaces could be further hardened to ensure availability when malicious code controls the phone. On the whole, remote management can provoke a concern on user privacy but we do not consider an insider attack which is taken by authorized employees (e.g., a patch developer) maliciously inject an infected patch through a secured provisioning channel. Protection against unauthorized modification of patch can be achieved by "message authentication code (MAC)" or "digital signature".

Given an over-the-air provisioning architecture, possible responses to an intrusion event include:

1. Sending a warning to users of the mobile nodes
2. Deactivating mobile nodes
3. Disable the short-range network interface on mobile nodes
4. Installing port or content-based filters
5. Installing patches to remove exploits
6. Provisioning patches to remove the worm

All of these responses can slow or stop the spread of the virus, however, they also incur user inconveniences of its own. For example, frequent use of response 1 may reduce its effectiveness, response 2 may prevent emergency calls, and response 3 may prevent the use of hands-free operation by drivers. Responses 3-6 require a more detailed understanding of the worm implementation and so may allow the worm to spread unrestricted for a period of hours or days. Even then, installing hastily developed patches often leads to failures on a subset of phones.

We define the *intrusion response planning problem* as identifying an optimal set of nodes to minimize the impact of the worm *and* the inconvenience and dangers cause by (partial) service outages due to the response. An optimal response plan only targets nodes that have already been infected or will be infected until the provisioning process is completed.

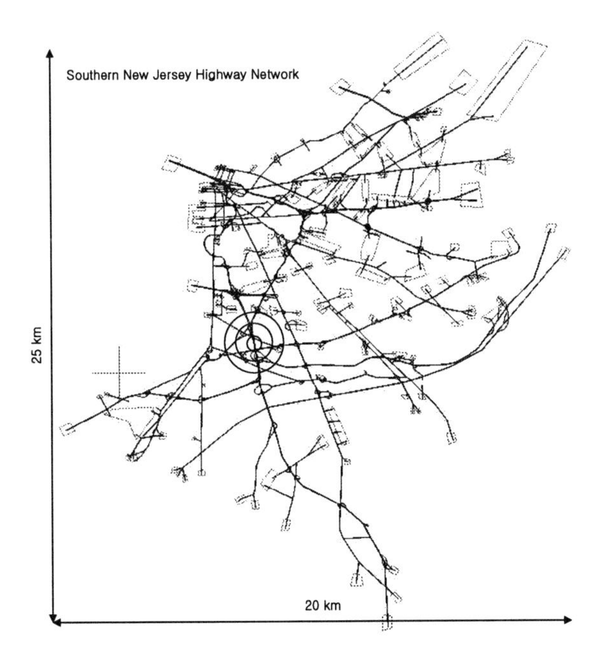

Fig. 7.1. Southern New Jersey highway network modeled in PARAMICS microscopic simulation software. It contains 2162 nodes, approximately 4000 links and 137 demand zones. Probe vehicles are selected randomly during the simulation process as they leave their respective origin zones. At each time step of the simulation (0.5 seconds), the x and y coordinates of the probe vehicles are recorded until they reach their destination zones. The location marked by the smallest circle indicates the position of initially infected vehicle.

7.2.2 Propagation case study in vehicular networks

To assess the threat posed by mobile worms in vehicular networks, we conduct an experiment which shows how fast mobile worms propagate (i.e., propagation speed) and how many vehicles can be infected over time (i.e., infection rate). In this case study, we consider the section of the southern New Jersey highway network shown in figure 7.1 and generate around 1800 vehicles corresponding to 5% of total traffic on peak hour. We create an initially infected node on the center of map and use a Susceptible-Infectious-Recovered (SIR) model for epidemic dynamics. We set the ad hoc communication range to 200 meters. Each vehicle's movement is modeled by a well-known microscopic traffic simulator, PARAMICS [18]. (For more details on the simulation model, see section 7.4.2.)

The case study in figure 7.2 shows that mobile worms can infect vehicles within an 11.6 kilometers radius within only 10 minutes. At this speed, mobile worms can traverse New Jersey from North to South in four hours (about 280 kilometers). Figure 7.3 shows that it takes about 13 minutes (800 seconds) to infect 90 percent of 1839 vehicles in southern New Jersey area.

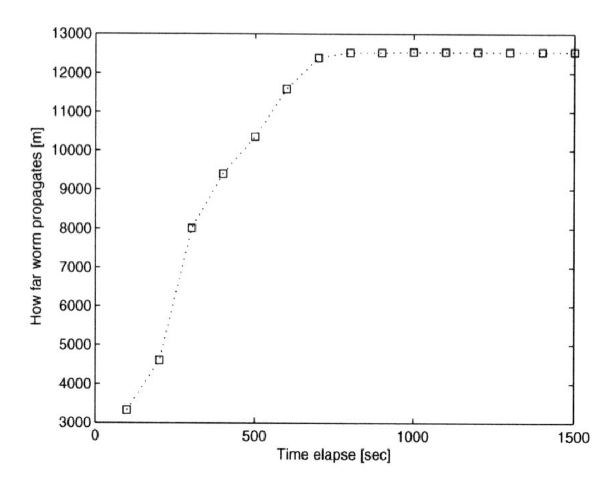

Fig. 7.2. The spatial propagation of mobile worms in southern New Jersey highway network. At each time unit, y value of each point depicts the Euclidean distance between the farthest infected vehicle and the origin of mobile worm.

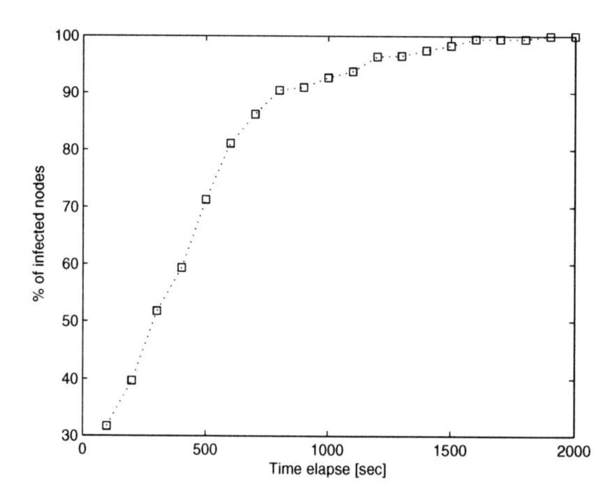

Fig. 7.3. The infection rate over time in southern New Jersey highway network. Totally, 1839 vehicles are injected onto map and 90 percent of them are infected within 800 seconds, approximately 13 minutes.

Staniford and Paxson [19] stated that conventional worms can infect up to 300,000 hosts within 8 hours and fast scanning worms such as flash worm can infect even faster (same number of hosts within 1 hour). Compared to Internet worms, mobile worms are slower but fast enough to make containment difficult.

7.3 Quarantine Boundary Estimation

The optimal response set can be best found through an estimation technique because the service provider's knowledge about the spread of the mobile worm is incomplete. Anomaly reports usually trickle in only after nodes are infected and may be severely delayed in areas of sparse coverage from the infrastructure wireless network.

7.3.1 A macroscopic model of worm propagation

Diffusion-reaction and advection models [20] have been successfully applied to describe the spatial and temporal distributions of diverse phenomena ranging from animal dispersion [5] to groundwater contamination.

The diffusion-reaction model comprises a diffusion process and a reproduction process. The diffusion process describes random movements and is characterized by the diffusion coefficient D. The reproduction process describes the exponential population growth and is specified by parameter α. Equation 1 specifies the diffusion-reaction model. It assumes polar coordinates centered at the position of an initially infected node (r indicates the distance from the origin), isotropic dispersal with constant diffusivity D, and growth proportional to the population density S.

$$\frac{\partial S}{\partial t} = \frac{D}{r}\frac{\partial}{\partial r}\left(r\frac{\partial S}{\partial r}\right) + \alpha S \tag{1}$$

This model has a closed form solution by solving under the initial condition that at time $t = 0$, m infected nodes are concentrated at location of *patient 0* ($r = 0$). From this solution shown in equation 2, the radius R of the frontal wave can be calculated from the propagation speed which depends on α and D as described in equation 3.

$$S = (m/4\pi Dt)\exp(\alpha t - r^2/4Dt) \tag{2}$$

$$R = 2\sqrt{\alpha D}t \tag{3}$$

Thus the propagation boundary is proportional to the time since the outbreak, t and the boundary moves with velocity $v = 2\sqrt{\alpha D}$. The parameter α and D are depended on the exact scenario. Table 7.1) identifies the parameter dependencies in an automotive scenario.

When a toxic pollutant diffuses going along the groundwater paths, its model consists of a uni-directional movement by mean flows, called advection together with diffusion-reaction processes [23]. In vehicular network, advection term is governed by the velocity u in x-axis and v in y-axis in two-dimensional space.

[5] An early notable application of diffusion-reaction model was designing a hostile barrier for stopping the dispersal of Muskrats. In 1905, Muskrat was imported to Europe but some of them escaped and started to reproduce in the wild [21]. Skellam [22] later modeled the dispersal of Muskrats though a diffusion-reaction equation.

Table 7.1. Mapping of model parameters to automotive networking scenario.

Model Parameter	Correspondence in automotive scenario
Diffusivity	Models minor roads and collector streets or pedestrian movements
Growth rate	Rate of new infections depends on density and distribution of susceptible nodes, communication range, and node velocity
Origin	Positions of initially infected nodes

If we take an advection effect and ignore a diffusion process, equation 1 is changed into an advection equation model described by equation 4.

$$\frac{\partial S}{\partial t} = -\frac{\partial}{\partial x}(uS) - \frac{\partial}{\partial y}(vS) + \alpha S \tag{4}$$

This model can be used in modeling the behavior of mobile worms in highway networks (e.g., Southern New Jersey Highway Networks).

7.3.2 Algorithms

Given an initial position of each infected node i, (x_i, y_i) for all i at time T_o, the algorithms should estimate the frontal wave of propagation at

$$T_c = T_o + T_\Delta,$$

where T_o is the time of outbreak and T_Δ means time delay. We can divide the problem into estimating the worm propagation velocity and estimating the spatial distribution.

However, in the vehicular scenario, every road segment may have a different propagation velocity because vehicle speeds and inter-vehicle distances differ. Figure 7.4 illustrates how the relationship between communication range and inter-vehicle distance affects propagation velocity. In the case (a) the inter-vehicle distance R is greater than the communication range C_r, so that an infected car cannot communicate with neighboring cars. Thus, the propagation velocity V' is solely determined by the vehicle speed V. In case (b) however, the communication range is greater than the inter-vehicle distance. Thus the worm can travel over the wireless medium to the foremost car in communication range in addition to the vehicle speed. If a worm manages n such hops per second, this leads to the following equation.

$$V' = \begin{cases} V + nR \lfloor \frac{C_r}{R} \rfloor & \text{if } R \leq C_r \\ V & \text{else} \end{cases}$$

Because a one hop communication can never go farther than C_r, an upper bound for V' can be obtained by substituting C_r for $R(C_r/R)$, yielding

$$V' = V + nC_r \tag{5}$$

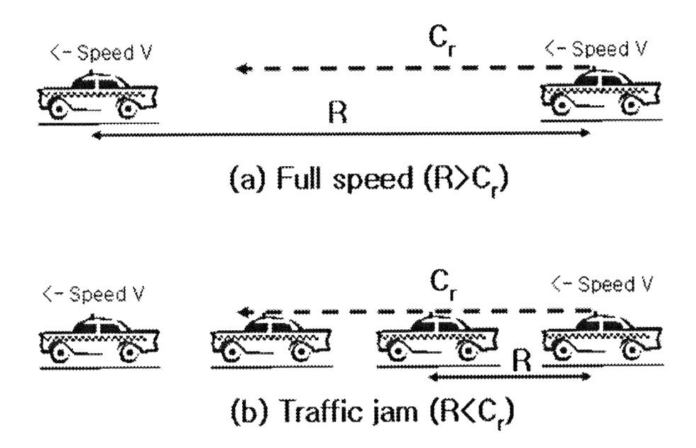

Fig. 7.4. Different proportions of inter-vehicle distance to communication range lead to different worm propagation velocities.

The inter-vehicle distance R and mean vehicle speed V on each highway segment can be obtained from Department of Transportation inductive loop sensors on an hourly basis, for example. They could also be inferred from tracking the position of probe vehicles on the highway network.

Given this propagation velocity, a straightforward isotropic estimate for worm distribution can be obtained with the diffusion-reaction equations. For each independent outbreak this approach yields a circular boundary estimate centered at the location of *patient 0* (at the time of the outbreak). The radius of the circle increases linearly with the time duration T_Δ since the outbreak.

This approach is suitable when nodes movements do not exhibit any directional trends, such as in a random walk. Estimation can be improved, however, when mobile nodes move on an underlying network of roads or walkways. We frame our discussion of this algorithm in the context of an automobile vehicular ad hoc network, but the concepts are generally applicable to nodes that follow a network of paths.

This algorithm assumes the availability of cartographic material so that the position of *patient 0* at the initial outbreak can be mapped onto a road segment. The maps must contain road classifications and the geographical positions of roads and their intersections. For example, this data is available from the US Geological Survey which publishes detailed transportation network information in the spatial data transfer standard. These maps also classify roads into expressways, arterial, and collector roads, according to their size and traffic volume. The algorithm also requires a mapping of the position of *patient 0* at the time of outbreak onto a road segment. This mapping can be achieved by finding the road segment with the minimum Euclidian distance to the *patient 0* position.

The key idea of this algorithm is to build an advection model using the transportation network information. The underlying heuristic is that the maximum propagation speed will be observed along the road network—propagation across parallel road segments in communication range and along smaller roads is ignored by this heuristic. The algorithm 1 follows all possible propagation paths using a traversal of the road network graph and a propagation speed estimate for each road segment. It outputs a polygon that includes all (partial) road segments that a worm could have reached in the time since the outbreak.

Algorithm 1 $QuarantineBoundaryEstimation$ generates a polygon which estimates the frontal wave of mobile worms at T_r given $Patient0$ at T_0.

1: {Inputs: $Patient0$, the position of initially infected node; T_0, the time of outbreak; T_c, the time of intrusion response; v_n, the average car speed on nth road segment; R_n, the average distance beween adjacent cars on nth road segment;
Parameters: J_n, nth junction's x and y coordinates and every junction should have information on its neighbor junctions; C_r, Communication range
Outputs: Quarantine polygons}
2: **(A) Estimate the worm propagation speed, V_n for all n with v_n and R_n**
3: **if** $R \geq C_r$ **then**
4: $V_n = v_n$
5: **else**
6: $V_n = v_n + \alpha * C_r$
7: **end if**
8: **(B) Estimate the spatial distribution**
9: Calculate $T_\Delta[0][0] = T_c - T_0$.
10: Locate the link (L_n) which $Patient0$ lies on.
11: Set $Patient0$ as the starting points of traversal and push it into queue, $Q[0]$
12: Keep pushing all junctions in two ways to be visited next in Q until the last level
13: $i = 0$;
14: **while** Any $T_\Delta[i][] \geq 0$ **do**
15: $i++$
16: K = the number of elements in $Q[i][]$
17: **for** $j = 1$ to K **do**
18: Save the parent junction of $Q[i][j]$ into $Prev$
19: $T_j = \frac{D(Prev,Q[i][j])}{V_n}$ where n is the link index between Prev and $Q[i][j]$
20: $T_\Delta[i][j] = T_\Delta[i-1][parent]$
21: **if** $T_\Delta[i][j] \geq T_j$ **then**
22: Generate a rectangular boundary from $Prev$ to $Q[i][j]$
23: **else**
24: Generate a rectangular boundary from $Prev$ to $T_\Delta[i][j] * V_n$
25: **end if**
26: $T_\Delta[i][j] = T_\Delta[i][j] - T_j$
27: **end for**
28: **end while**
29: Merge all rectangular boundaries into polygon.

For example, consider the section of the southern New Jersey highway network in figure 7.5. Assume that *patient 0* lies on the link L_n between junction 3 (J3) and junction 4 (J4). If we know the propagation speed V_n on that link, we can calculate

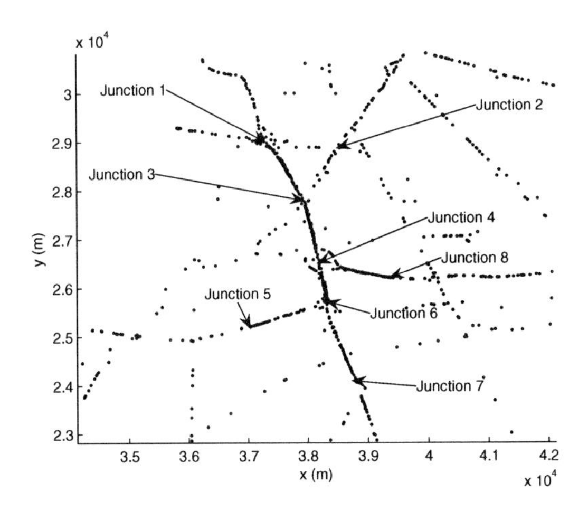

Fig. 7.5. In our target map, there are 8 junctions and 7 links between them. This region is the part of Southern New Jersey Highway Networks 7.1. Every black dot depicts the position of individual car at specific time.

after how much time a mobile worm arrives at either junction. Let us denote T_3 and T_4 for the arrival time at J3 and J4. If the time since outbreak

$$T_{\Delta} = T_c - T_o$$

is greater than T_3, the mobile worm has already passed this junction and has most likely propagated along both the link J1-J3 and the link J2-J3. This process is repeated for each link until a junction with arrival time greater than T_{Δ} is found. This segment is then only partially infected and the infection boundary is known based on the estimated link propagation speed. The same process is also repeated in the opposite direction from patient 0, towards J4. The algorithm then encloses each fully infected link in a rectangle with length and width set to the road length and road width, respectively. Partially infected links are only enclosed up to the infection boundary. All rectangles are then merged into a polygon.[6] Once we get a polygon, we group nodes within a polygon into the optimal response set by using 'Point-In-Polygon Algorithm [25]'.

7.4 Evaluation

This evaluation studies the performance of the quarantine boundary estimation algorithms in a random walk and a vehicular ad hoc network scenario. We compare the accuracy of the macroscopic quarantine boundaries against infection patterns generated by a microscopic simulation model.

[6] This can be implemented using well-known algorithms such as provided by the *polybool* function [24] in MATLAB.

7.4.1 Metrics and measures

Informally, the algorithm should maximize the number of infected nodes within the boundary and minimize the number of clean (uninfected) nodes within it. We measure the accuracy of the quarantine boundary estimation through detection and false-alarm probability.

The *detection probability* is defined as the ratio of infected nodes within the boundary to all infected nodes. More formally, $P_d = \frac{i}{I}$, where P_d is the detection probability, i is the number of infected nodes within the boundary and I is the total number of infected nodes. We define the *false-alarm probability* as the ratio of clean nodes within the boundary to all clean nodes. Accordingly, $P_f = \frac{c}{i+c}$, where P_f is the false alarm probability, c is the number of clean nodes within the boundary and C is the total number of clean nodes. Notice that $c + i$ is the number of nodes within the quarantine boundary and $C + I$ is the total number of nodes in the scenario. A perfect quarantine boundary has a detection probability of 1 and a false-alarm probability of 0.

The *Jaccard similarity* J provides a convenient way to combine above two probabilities into one number as an *ROC curve* (i.e., receiver-operating characteristics) does in detection theory community. It is defined as shown in equation (6), where X is the optimum quarantine boundary in x-y coordinates and Y indicates an estimated quarantine boundary.

$$J = \frac{2\,(|X \bigcap Y|)}{|X| + |Y|} \tag{6}$$

It can be computed from detection and false alarm probabilities by substituting $X = I$ and $Y = i + c$, yielding equation (7).

$$J = \frac{2P_d(1 - P_f)}{1 + P_d - P_f} \tag{7}$$

The Jaccard similarity lies in the interval $[0, 1]$ with 1 indicating a perfect estimate, corresponding to detection probability 1 and false-alarm probability 0. Jaccard similarity can be used to balance between detection probability and false alarm probability.

7.4.2 Simulation model

We use the SIR model [26] for implementing the dynamics among susceptible nodes, infected nodes and recovered nodes. This model is characterized by the fraction of nodes that are susceptible to infection, the infection probability when a susceptible node is in contact with an infected node, and a recovery probability. In our model a susceptible node is in contact with an infected node, if they are in communication range C_r of each other.

Generally, we chose aggressive parameters for our simulations to evaluate a near worst-case worm. We set the infection probability to 1, which assumes the absence of any communication errors. In other words if a susceptible node is within the communication range of an infected node it becomes infected. We assume that infected nodes can only be recovered by the service provider only if they are within the quarantine boundary. Worm propagation then depends on the communication range and the exact mobility model.

We choose the initially infected nodes randomly among all nodes in random walk scenario. However, in VANET scenario, we choose them only on the link between J3 and J4, which is at the center of the map in figure 7.5. The position of initially infected node is independent from the performance of our quarantine boundary algorithm, but placing them on that link enables us to extend the simulation duration.

For a random walk scenario, we choose 5 seconds as T_Δ. After T_Δ elapsed in pedestrian scenario, the number of infected nodes amounts up to 40-50% of whole nodes and the propagation for each initially infected node covers up to the circle with about 13m radius. Because our network is 50m by 50m, this amount of T_Δ is appropriate to measure detection, false alarm probabilities. In VANET case, we choose a time delay, T_Δ from 25 seconds to 45 seconds. In the case of T_Δ =45 seconds, the propagation approaches almost 5 links out of all 7 links.

For the *random walk model*, we chose parameters to reflect dense pedestrian movements with short-range (e.g., Bluetooth) communications. Node density is varied from 100 to 300 in a 50m by 50m area with node velocity ranging between 1m/s to 3m/s. Communication range is set to 5m, 10m, and 20m, to represent different path loss and interference environments.[7]

For the *vehicular scenario*, we obtained location traces from a microscopic traffic model for the PARAMICS transportation system simulator [18]. The model is calibrated to real traffic observed in a section of the southern New Jersey highway network. [27] The full simulation model contains 2162 nodes, approximately 4000 links and 137 demand zones, from which serve as origins and destinations for vehicles. Out of all vehicles in the simulation model a fraction of susceptible vehicles are selected randomly during the simulation process as they leave their respective origin zones. This ensures that the overall traffic patterns remain realistic even though we assume that only a percentage of cars is equipped with susceptible communications equipment. At each time step of the simulation (0.5 seconds), the x and y coordinates of the susceptible vehicles are recorded until they reach their destination zones. For a low susceptibility scenario we selected 200 vehicles and for a moderate susceptibility scenario we chose about 1800 random cars. This represents about 5% of total traffic during the simulation which was restricted to 4min 10s, for computational tractability. The communication range is set to 50m, 100m and 200m in this scenario. 200m approximates free space propagation of a DSRC system [28,29], while the shorter ranges model higher path loss environments, such as in congested traffic.

[7] These parameters approximate a sport event environment such as the one in the Helsinki Olympic Stadium, where an outbreak of the Cabir virus was reported [6].

7.4.3 Pedestrian scenario results

To gain a better understanding of the effect of different model parameters we first discuss results from the less complex diffusion-reaction estimation model. The estimator's worm propagation speed is set to 2.56 m/s and the time delay T_{Δ} is set to 5 seconds for these experiments.

Figure 7.6 shows estimation accuracy of the diffusion-reaction estimator for different node densities. Mean and standard deviation for one hundred trials are shown. A mean detection probability between 95%-100% can be achieved with a false alarm rate of approximately 40%-50%. Our quarantine method behaves slightly more effective in the 200 node network because the worm propagation speed best matched this case. A change of +/-100 nodes increases the false alarm probability by about 10%.

The following results analyze the worm propagation speed in more detail. The speed is affected by node density, communication range, and node mobility. Figure 7.7 shows the distance of the farthest infected node from original position of *patient 0* over differnt node velocities. Node density is set to 200 in the 50m by 50m region and communication range is 10m. Again, the graph shows mean and standard deviation over one hundred trials. As expected, propagation speed increases with node velocity. An increase in node velocity has an additive effect on propagation speed. The graph also exposes that propagation speed remains constant over time, further supporting that a linear model fits well. A linear regression for v=2m/s yields intercept 2.1 and slope 2.8m/s.

The effect of changes in communication range C_r to worm propagation speed are shown in figure 7.8. Node velocity is set to 1m/s and other parameters remain the same as before. Propagation speed increases with higher node velocity. A larger communication range increases the likelihood that susceptible nodes are in rage, which hastens the spread of the worm. Propagation speed remains near-constant over time for each communication range.

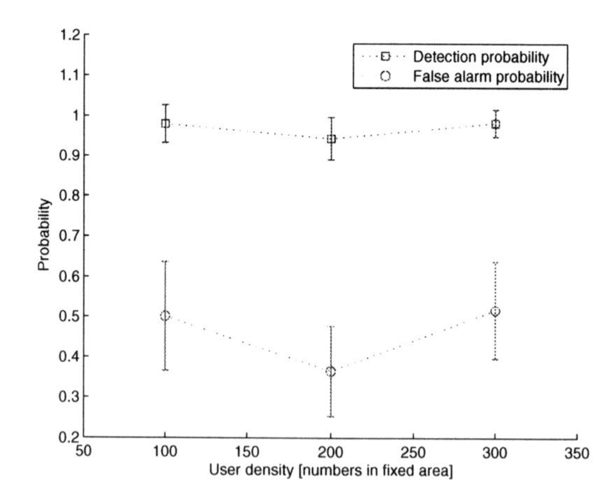

Fig. 7.6. Estimation accuracy of diffusion-reaction model for random-walk scenario.

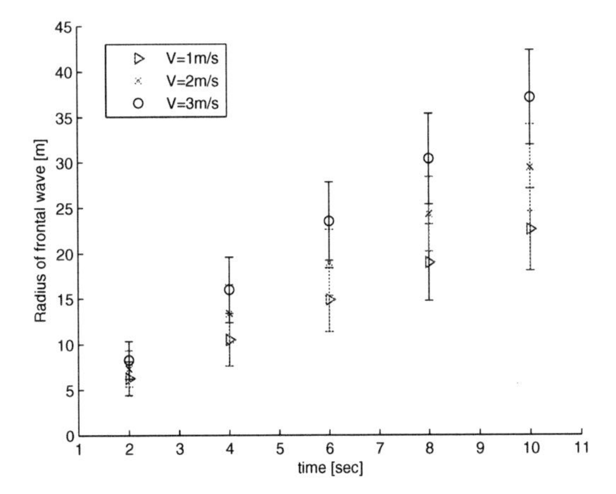

Fig. 7.7. Distance of the farthest infected node from the outbreak position over time. Increasing node velocity has an additive effect on propagation speed. Propagation speed remains constant over time.

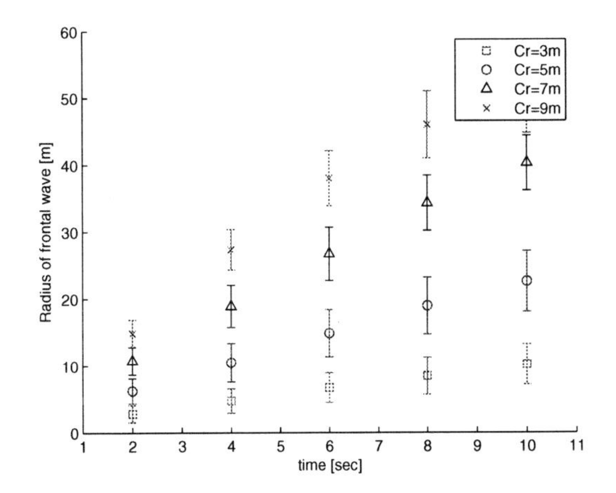

Fig. 7.8. Dependency of propagation speed on communication range C_r. A larger communication range increases the likelihood that susceptible nodes are in rage, which hastens the spread of the worm.

7.4.4 Vehicular scenario results

The first experiment measures the worm propagation velocity that can be expected in a highway outbreak. While prior works [30–32] have developed analytical equations for information propagation speed on road networks, these are not easily transferable to the worm scenario. The average radius of frontal wave is estimated by averaging

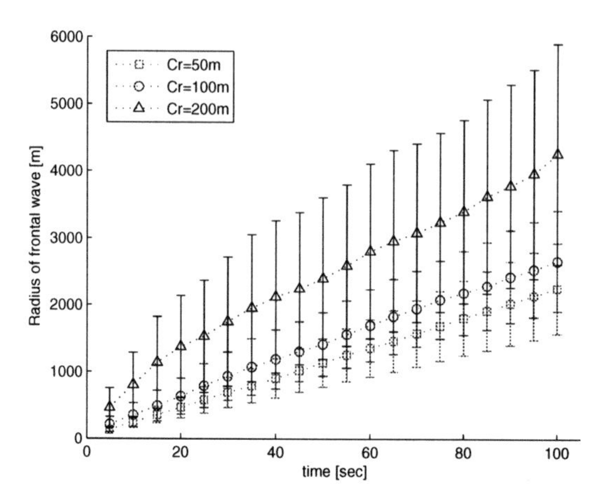

Fig. 7.9. Worm propagation in highway model with 5% of vehicles susceptible.

50 simulations and it is repeated for different communication ranges (50m, 100m and 200m). The estimated radius of frontal wave is shown in figure 7.9. The results show that for a communication range of 200m, the worm travels at a mean velocity of about 75m/s, significantly faster than typical highway traffic. Lower communication ranges result in reduced velocity.

The next experiment compares the estimation accuracy of the advection model over the diffusion-reaction model in the highway scenario. The communication range is set to 100m. Figure 7.10 and figure 7.11 show the detection and false alarm probability, respectively. The results from the advection algorithm described in section

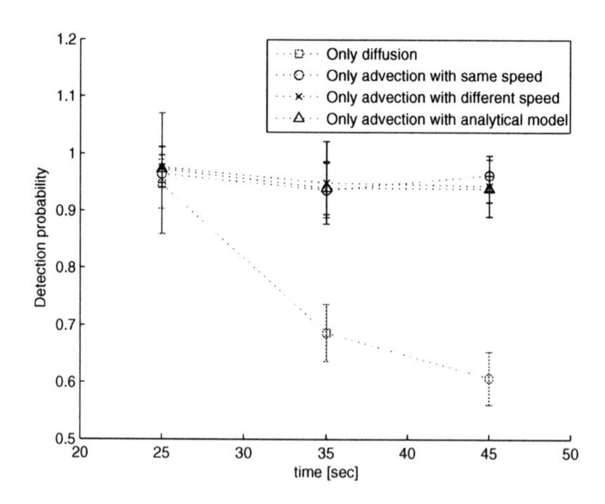

Fig. 7.10. Detection probability on highway network. The advection models achieve superior accuracy over the diffusion-reaction model.

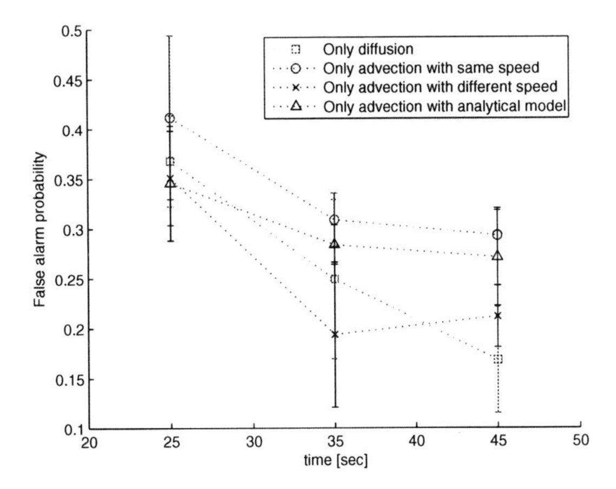

Fig. 7.11. False-alarm probability on highway network. The advection model's better detection probability does not lead to a significant increase in false alarms.

7.3 are labeled "advection with analytical model". To allow a more detailed analysis, the graphs also contain two additional curves, which assume that a more precise estimate of worm propagation speed is available. In the "advection with same speed" approach, we use the average worm propagation speed (obtained from the previously described simulation) for all road segments. The "advection with different speed" approach, uses more detailed speed estimates, one per road segment, also derived from simulations.

These figures show that the advection models achieve superior detection probability over the diffusion-reaction model, while the false-alarm probability does not differ more than about 10% between advection and diffusion. The detailed knowledge about information propagation speed does not lead to a discernible improvement in detection probability. However, when worm propagation speed is known per road segment, the mean false alarm probability improves by up to 10%. This shows that at least slight improvements to the presented estimation techniques are possible.

7.5 Discussion

Location-based quarantine boundary estimation is achieved in two steps: (1) locating *patient 0* and (2) estimating a quarantine boundary based on *patient 0* location and propagation speed. Thus the quarantine boundary estimation depends on accurate knowledge of patient 0 location. So far we assumed that the service provider can locate *patient 0* accurately from a set of intrusion reports. Here, we discuss how the location might be obtained, if initially unknown. We leave the detailed analysis for future work. We also discuss the the impact of slightly inaccurate quarantine boundaries and other synergies between computer security and ecology.

7.5.1 Estimating patient 0 location

In a pedestrian scenario, triangularization can help a service provider locate the initially infected node. We assume that only a limited number of mobile units have intrusion detection systems due to high cost. If a mobile worm originates from point (x_0, y_0) at time t_0 and propagates isotropically with speed v in two dimensional space, a distributed intrusion detection system at (x_i, y_i) eventually reports an anomaly at time t_i to the service provider. Every IDS report forms a nonlinear equation expressing that the mobile worm can propagate from (x_i, y_i) to (x_0, y_0) within

$$t_\Delta = t_i - t_0$$

at the speed of v. Assuming prior knowledge of propagation speed and more than three intrusion reports, the service provider can apply triangularization algorithms [33] (similar to the GPS localization problem). Without this prior knowledge numerical methods such as Newton-Raphson could be applied, but at a higher computational cost.

Because the vehicle scenario confines mobile worm propagation to the road network topology rather than an isotropic two-dimensional space, it requires a more complex solution with three steps: (1) guessing the approximate road segment on which the *patient 0* location lies, (2) setting up and solving a set of linear equations using recursive least squares (RLS), and (3) repeating the second step over neighboring segments around the starting segment. Given at least three reports, triangularization might be used to obtain the approximate road segment. The second step refines the estimated *patient 0* position within the approximate segment given from the previous step using linear equations where the unknown variables are time t_0 and the relative position on the given road segment. After repeating this step for neighboring segments, the segment with the best least squares fit is chosen.

7.5.2 Effectiveness of partial containment

Estimation will necessarily lead to imperfect containment. Can this effectively slow worm propagation? We model the accuracy of quarantine boundary through an immunization probability P_{imm} between 0.8 and 1 and simulate worm propagation in the pedestrian random-walk scenario after such an imperfect containment. Figure 7.12 depicts the infection rates after one containment was performed at $T_c = 5seconds$. Detection probabilities greater than 0.95%, such as achieved by the advection model, significantly slow the propagation of a worm, yielding additional analysis time for security engineers.

So far, we assumed that the intrusion response is only performed once. Repeated application, however, could further slow worm propagation. One approach would be to wait for any intrusion reports after the first response and then retry with an enlarged boundary. Another approach would treat every remaining infectious node as a new outbreak. However, this requires changes to the estimation model because the worm will continue to spread from multiple locations, rather than a single origin.

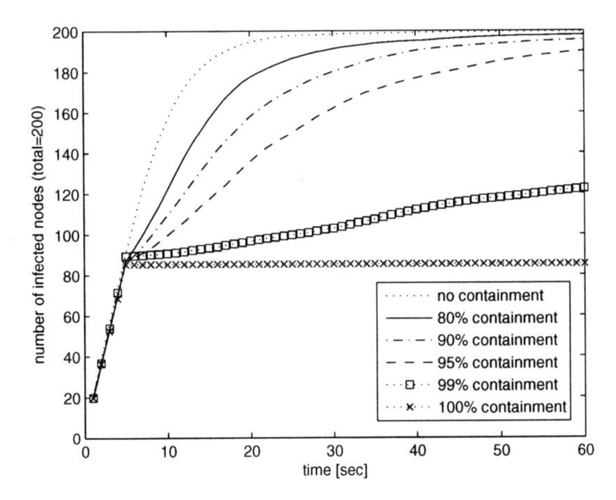

Fig. 7.12. Effect of imperfect containment on worm propagation speed. Containment techniques with more than 95% detection probability can significantly slow worms.

The current solution aims for a high detection probability, to effectively slow worms. In some scenarios a more balanced approach that also minimizes the false alarm probability may be desirable. Higher Jaccard similarity values, for example, can be obtained when small reductions in detection probability yield large reductions in false-alarm probability. To optimize Jaccard similarity we could choose a smaller radius $\hat{R} = \gamma R = \gamma 2\sqrt{\alpha D t}$ for the random walk scenario (γ is less than 1). $\hat{R}$ denotes the effective radius which equals the square root of the propagation area enclosed by a real boundary (not a circle) against time. Our usage of R instead of $\hat{R}$ also explains the adaptation of our algorithm over different node densities.

7.5.3 Other synergies between ecology and computer security

The successful application of ecological models to estimating worm propagation raises the question about other potential synergies between the fields. Biologically inspired interdisciplinary work has long affected computer security. For example, computer immunology improves virus defenses [34]. Epidemiology enables us to investigate the spread of computer viruses on a hybrid networks that combine computer network and social networks, such as email [35]. In ecology the Allee effect (or reduced per capita reproduction when animals are scarce) may be useful for describing the dynamic change of the infection rate when we have disconnections in the ad hoc network. The effect of dispersal on competing populations (e.g., Predator-Prey model) also holds promise for modeling competition[8] or the cooperation of malicious codes [10].

[8] In 2001, the counterattacking CodeGreen appeared to disinfect CodeRed.

7.6 Related Work

Moore and colleagues [36] investigated and compared the existing containment methods for Internet worms which can be implemented in gateway, firewall and router. The hierarchical structure of the Internet allows an administrator to partition and shut down a local sub-network which is infected. In wireless networks, however, an infected node can move and communicate with a susceptible node via localized interaction such as Bluetooth. Our work instead focuses on estimating the geographic propagation pattern of short-range wireless worms. The notion of locality is less meaningful in wired networks where worms often use random probing.

Khayam and Radha [37] investigated the parameters governing the spread of active worms over VANET. They define the average degree of a VANET node and use a SIR model for the spread of worms. In our work, we provide a spatial and temporal distribution of the propagating worms rather than an infection rate over time. Bose and Shin [5] explored the impact of the mobile worm propagating through both infrastructure (e.g., MMS/SMS) and local interaction (e.g., bluetooth) on the cellular networks and showed that the existence of the latter expedites mobile worm propagation. The movement patterns are simulated by 2-dimensional random waypoint and Gaussian-Markov models with relatively low speed. Also, the investigation concentrates on the number of infected nodes without considering spatial distribution. Further they developed a server-based intrusion detection system to automatically identify and contain the intrusion of malware propagating via SMS/MMS in cellular networks in their recent work [38]. However, their intrusion detection algorithm does not consider the propagation via local interaction. Thus, it leaves room for being merged with a location-based quarantine estimation for the attack of unknown malwares using both of two infection vectors (i.e., infrastructure-based messaging and local interaction).

Wu and Fujimoto [30] presented an analytical model for information propagation in Vehicle-to-Vehicle Networks. Worm propagation is very similar to information dissemination except that it has an malicious purpose and it lacks cooperation of neighboring nodes. Our work concentrated on practical estimation algorithms that are tractable for larger highway networks. We also presented simulation results from a calibrated highway simulation.

Several intrusion detection system for wireless ad hoc networks have been designed [14,39]. Zhang and Lee present a collaborative intrusion detection system for ad hoc and assume that every node runs an IDS agent. Anjum and colleagues have investigated the optimal placement of intrusion detection nodes in an ad hoc network to reduce the need for one IDS agent per node [15]. This intrusion detection work concentrates mostly on external attacks such as distributing erroneous routing information. They do not address how to catch up with a propagating worm. Our work shows how to take advantage of a wireless infrastructure network and how to forecast the propagation of the worm.

The spatial and temporal modeling of tumor [40] or biological invasion has common factors with mobile worm propagation research. Shigesada et al. [41] modeled the spatial expansions of biological invasion and tumor invasion by a stratified dispersal process in which a combination of neighborhood diffusion and long-distance dispersal is expressed in a differential equation form. Its early expansion mainly occurs by neighborhood interaction, but later new colonies are created by long-jump migrants which accelerate the expansion.

7.7 Conclusions

Wireless ad hoc networks requires a new worm intrusion response architecture and mechanisms because it lacks central infrastructure choke-points such as routers, gateways and firewalls where network intrusion detection such as address blacklisting or content filtering can take place. We have considered a scenario in which a service provider manages the security of an hybrid (ad hoc with wide-area network) network over a low-bandwidth, wide-area infrastructure wireless network. This work proposed to develop location-based quarantine boundary estimation techniques. These techniques let service providers identify the current set of likely infected nodes when intrusion information is incomplete or delayed. Specifically, we found that

- a mobile worm could spread in a typical highway network with a mean velocity of about 75m/s even though only 5% of vehicles are susceptible to attack.
- advection-based estimation techniques can estimate the group of currently infected nodes with a detection probability greater than 95% and a false-alarm rate of less than about 35%. This provides a significant improvement over having to target a response at all nodes in a large geographic region.

There are several directions for future work. First, the algorithm should be more robust to the inaccuracy of geographic origin of the outbreak. Second, it appears valuable to develop techniques that effectively address partial outages of the wide-area wireless network. Third, the system could take advantage of propagation speed information gained from the time difference in intrusion reports from different nodes. Fourth, we could borrow a stratified dispersal process [41] from a biological invasion to model the mobile worm propagation which also uses infra-structure network such as MMS or SMS-based downloaders (long-distance dispersion) as well as a local interaction (neighborhood diffusion). Fifth and finally, location privacy preservation should be addressed in the design of quarantine boundary estimation algorithm. The location information in intrusion/anomaly reports could enable the authority to locate or track users.

Acknowledgment

The authors would like to thank Dr. Ozbay for providing a location trace file from Southern New Jersey Highway Network available for the purposes of this study.

References

1. DSRC-5GHz-Standards-Group. *Standard Specification for Telecommunications and Information Exchange Between Roadside and Vehicle Systems - 5GHz Band Dedicated Short Range Communications (DSRC) Medium Access Control (MAC) and Physical Layer (PHY) Specifications.* ASTM E2213-03, 2003.
2. Qing Xu, Raja Sengupta, and Daniel Jiang. Design and analysis of highway safety communication protocol in 5.9 ghz dedicated short range communication spectrum. In *IEEE VTC Spring 2003*, April 2003.
3. Jijun Yin, Tamer ElBatt, Gavin Yeung, Bo Ryu, Stephen Habermas, Hariharan Krishnan, and Timothy Talty. Performance evaluation of safety applications over dsrc vehicular ad hoc networks. In *VANET '04: Proceedings of the 1st ACM international workshop on Vehicular ad hoc networks*, pages 1–9, 2004.
4. Jing Su, Kelvin K. W. Chan, Andrew G. Miklas, Kenneth Po, Ali Akhavan, Stefan Saroiu, Eyal de Lara, and Ashvin Goel. A preliminary investigation of worm infections in a bluetooth environment. In *WORM '06: Proceedings of the 4th ACM workshop on Recurring malcode*, pages 9–16, New York, NY, USA, 2006. ACM Press.
5. Abhijit Bose and Kang G. Shin. On mobile viruses exploiting messaging and bluetooth services. In *Proceedings of IEEE/Create-Net SecureComm 2006*, Baltimore, MD, USA, August 2006.
6. Reuters. Mobile phone virus infects helsinki championships: The cabir virus uses bluetooth to jump between cell phones. `http://www.computerworld.com/securitytopics/security/virus/story/0,10801,103835,00.html`, Aug 2005.
7. ZDNet-UK. Year-old bluetooth vulnerability invites mobile worm. `http://news.zdnet.co.uk/internet/security/0,39020375,39162400,00.htm`, Aug 2004.
8. Xatrix-Security. Widcomm bluetooth connectivity software multiple buffer overflow vulnerabilities. `http://www.xatrix.org/article.php?s=3663`, Aug 2004.
9. David Dagon, Tom Martin, and Thad Starner. Mobile phones as computing devices: The viruses are coming! *IEEE Pervasive Computing*, 3(4):11–15, 2004.
10. Peter Szor. *The Art of Computer Virus Research and Defense.* Addison-Wesley Professional, symantec press, 2005.
11. Grant A. Jacoby, Randy Marchany, and Nathaniel J. Davis IV. Using battery constraints within mobile hosts to improve network security. *IEEE Security and Privacy*, 4(5):40–49, 2006.
12. Ross J. Anderson. *Security Engineering: A Guide to Building Dependable Distributed Systems.* Wiley, January 2001.
13. Thomas Martin, Michael Hsiao, Dong Ha, and Jayan Krishnaswami. Denial-of-service attacks on battery-powered mobile computers. *percom*, 00:309, 2004.
14. Yongguang Zhang and Wenke Lee. Intrusion detection in wireless ad-hoc networks. In *The Sixth International Conference on Mobile Computing and Networking (MobiCom).* ACM, August 2000.
15. Farooq Anjum, Dhanant Subhadrabandhu, and Saswati Sarkar. Intrusion detection for wireless adhoc networks. In *Proceedings of Vehicular Technology Conference, Wireless Security Symposium.* IEEE, October 2003.
16. Open-Interface. Bluetooth security overview. `http://www.oi-us.com/service_additions/security_whitepaper_docpage.html`, Dec 2005.

17. Open-Mobile-Alliance. Provisioning architecture overview. http://www.openmobilealliance.org/release_program/docs/ClientProv/ V1_1-20050428-C/OMA-WAP-ProvArch-v1_1-20050428-C.pdf, Apr 2005.
18. Quadstone-Limited. Paramics v4.0 - microscopic traffic simulation system. www.paramics-online.com.
19. Stuart Staniford, Vern Paxson, and Nicholas Weaver. How to own the internet in your spare time. In *Proceedings of the 11th USENIX Security Symposium*, pages 149–167, Berkeley, CA, USA, 2002. USENIX Association.
20. Akira Okubo and Simon A. Levin. *Diffusion and Ecological Problems: Modern Perspectives*. Springer, 2002.
21. C. S. Elton. *The Ecology of Invasions by Animals and Plants*. Methuen Co. Ltd., London, 1958.
22. J. G. Skellam. Random dispersal in theoretical populations. *Biometrika*, 38(4):196–218, 1951.
23. Muhammad Sadiq. *Toxic metal chemistry in marine environments*. CRC, New York, 1992.
24. MathWorks-Inc. Overlaying polygons with set logic. http://www.mathworks.de/access/helpdesk/help/toolbox/map/polybool.html, 2005.
25. Darel R. Finley. Point-in-polygon algorithm: Determining whether a point is inside a complex polygon. http://www.alienryderflex.com/polygon/, 1998.
26. N. T. Bailey. *The Mathematical Theory of Infectious Diseases and its Applications*. Hafner Press, New York, 1975.
27. K. Ozbay and B. Bartin. South jersey real-time motorist information system. NJDOT Project Report, March 2003.
28. J. P. Hubaux, S. Capkun, and J. Luo. The security and privacy of smart vehicles. *IEEE Security and Privacy Magazine*, 2(3):49–55, June 2004.
29. M. Raya and J. P. Hubaux. The security of vehicular ad hoc networks. In *Proceedings of SASN'05*, November 2005.
30. H. Wu, R. Fujimoto, and G. Riley. Analytical models for data dissemination in vehicle-to-vehicle networks. In *Proceedings of IEEE 2004-fall Vehicle Technology Conference (VTC)*, September 2004.
31. Linda Briesemeister, Lorenz Schafers, and Gunter Hommel. Disseminating messages among highly mobile hosts based on inter-vehicle communication. In *IEEE Intelligent Vehicles Symposium*, October 2000.
32. Samir Goel, Tomasz Imielinski, and Kaan Ozbay. Ascertaining the viability of wifi based vehicle-to-vehicle network for traffic information dissemination. In *Proceedings of the 7th Annual IEEE Intelligent Transportation Systems Conference (ITSC)*, October 2004.
33. Zang Li, Wade Trappe, Yanyong Zhang, and Badri Nath. Robust statistical methods for securing wireless localization in sensor networks. In *IPSN*, pages 91–98, 2005.
34. S. Forrest, S. Hofmeyr, and A. Somayaji. Computer immunology. *Communications of the ACM*, 40(10):88–96, 1997.
35. M. E. J. Newman, S. Forrest, and J. Balthrop. Email networks and the spread of computer viruses. *Physical Review*, 66(035101), 2002.
36. David Moore, Colleen Shannon, Geoffrey M. Voelker, and Stefan Savage. Internet quarantine: Requirements for containing self-propagating code. In *INFOCOM*. ACM, 2003.
37. Syed A. Khayam and Hayder Radha. Analyzing the spread of active worms over vanet. In *Proceedings of the first ACM workshop on Vehicular ad hoc networks*, January 2004.
38. Abhijit Bose and Kang G. Shin. Proactive security for mobile messaging networks. In *WiSe '06: Proceedings of the 5th ACM workshop on Wireless security*, pages 95–104, New York, NY, USA, 2006. ACM Press.

39. A. Mishra, K. Nadkarni, and A. Patcha. Intrusion detection in wireless ad hoc networks. *IEEE Wireless Communications*, 11:48–60, 2004.
40. K. Iwata, K. Kawasaki, and N. Shigesada. A dynamical model for the growth and size distribution of multiple metastatic tumors. *Journal of theoretical biology*, 203(2): 177–186, 2000.
41. N. Shigesada, K. Kawasaki, and Y. Takeda. Modeling stratified diffusion in biological invasions. *American Naturalist*, 146(2):229–251, 1995.

8 Approaches for Ensuring Security and Privacy in Unplanned Ubiquitous Computing Interactions

V. Ramakrishna, Kevin Eustice, and Matthew Schnaider

Laboratory for Advanced Systems Research, Computer Science Department, University of California, Los Angeles, CA 90095.

8.1 Introduction

Ubiquitous computing promises a vision of computing capabilities at any place and at any time, supporting all kinds of human activities, including even the most mundane. A transition from mobile computing to ubiquitous computing is well underway thanks to both academic research efforts and commercial enterprises. Three important technological factors are contributing to this transition: 1) rapid growth and proliferation of wireless networking facilities, 2) computing and sensing components embedded in our surrounding environments, and 3) availability of smaller portable devices that can run most applications required by a mobile user. Mark Weiser envisioned a future in which computers would fade into the background [33]. A more realistic vision, and one that is currently attainable, still involves devices that are recognizable to users as computers. This model of computing is typically distinguished from ubiquitous computing (*ubicomp*) as *pervasive computing*. In the pervasive computing paradigm, devices and networks communicate with each other and deal with each other in a more aware and intelligent fashion, without involving a human unless absolutely necessary. Most of these interactions occur in a mobile context and in an unplanned fashion. The onus is upon the devices and the applications to ensure that tasks proceed smoothly, hiding details from users. The challenges in pervasive and ubiquitous computing are similar to mobile computing, but with a higher scale of mobility, dynamism, and heterogeneity.

Primary networking challenges have more or less been addressed. These include the ability to discover networks and associate with them, and the addressing issues that are necessary to establish and maintain network connections. Efforts at the application layer have been made, and are still ongoing, to achieve seamless mobility of networked applications. As a result, the networking infrastructure can now handle complex tasks that were formerly relegated to the user.

Even as we design technology with new and better functionality, we must explore potential pitfalls. One or more of the participants in a mobile interaction may not play by the rules the designers of the mechanisms envisioned. Attackers could use their anonymity and the nature of network-based protocols to breach the security of trusting devices or obtain sensitive information. The networking infrastructure that makes mobile computing possible could also be subverted for illegitimate purposes. We will further explore the vulnerabilities inherent in these unplanned interactions and discuss how a complex balancing act is required to make ubiquitous computing usable, as well as secure.

8.1.1 Characteristics of ubiquitous computing interactions

Ubiquitous interactions rely primarily on wireless network connectivity between numerous classes of devices. In this context, wired portable computing is significantly less interesting, and the networking and addressing issues have, for the most part, been dealt with; additionally, there is a much higher level of trust and accountability.

Interactions among mobile devices and ubiquitous infrastructure components are directed towards the discovery and access of external resources and information that are required for local applications. These include services provided by the immediate environment—typically wireless connectivity, connections to remote computers through the Internet, and sensory output. Most current applications of mobile computing involve access of web-based services. This requires that devices be able to associate with networks and configure Internet connections; the remaining application tasks are explicitly performed by the users. The transformation to a pervasive computing environment will increase the demands on the devices and the networks to which they connect. A much wider variety of tasks will be supported, and the devices must be more intelligent and aware in order to minimize the work that users must do. Users will expect less intrusiveness, seamless communication, and better performance.

Devices and networks will become more autonomic, specifically more self-configuring, self-adjusting, and self-healing. In the simplest form of mobile computing, where users explicitly handle applications and provide other input, the networking issues have relatively fewer security implications. When devices and applications are expected to perform tasks that satisfy user desires, without low-level user input, and sense and adapt to context changes, the security problems are magnified. Workable solutions must be provided so that users can trust their devices to run in an automated fashion and handle private data.

Ad hoc or unplanned interactions, which we believe will be very common in the emerging computing landscape, will present situations where there is a lack of familiarity or trust among the interacting entities. We cannot guarantee that different mobile devices and networks will have the same security or data privacy standards, and one challenge is to determine the opposite party's standards. Even in cases where interactions occur between known entities or entities with verifiable security relationships, the lack of trustworthiness of the wireless communication medium calls for precautions. This medium enables anonymity of entities; if such entities turn out to be malicious or compromised, they could provide fake services and obtain sensitive information. It is conceivable that the problem could be mitigated somewhat through the imposition of strict security standards and a universal trust framework, but such a worldwide standard would be impractical and impossible to enforce. It would also limit the options for each independent domain to determine its security policies. It also does not solve the problem of adaptation with context, since all possible situations cannot be planned for in advance.

8.1.2 Trading off security, privacy and usability

Security has proven to be a challenge when it conflicts with user convenience and ease of use. Users dislike entering passwords repeatedly in order to perform tasks that require extra privilege. If the system provides an option of storing the password for subsequent use, many users would make use of it. Likewise, when a sensitive transaction requires the release of identity information and secret keys, privacy is often sacrificed with little thought. These examples and others indicate that there is a three-way tradeoff in security, privacy and usability that every system designer must address. In this context, we define usability as the ease of handling devices and applications, with minimal input and feedback required from the user for successful operation.

This complex tradeoff acquires a new dimension in mobile and ubiquitous computing due to the wireless medium, the open environments, the unplanned nature of interactions, and the anonymity of computing entities. In a static context, there is an added degree of trust, which is absent in a mobile wireless context. When communicating with strangers, the more knowledge a device gains about the other party, the better it can assess the appropriate level of trust to place in that party. Intrusive procedures for assessing trust could be used, indirectly leading to more security. This would make an entity more confident about allowing access to a local resource or giving up some private information in the hope that this might result in some benefit without the cost of misuse. Trust-based security therefore inevitably results in a loss of privacy. Conversely, a conservative policy could result in more privacy but a lower probability of a successful interaction because neither entity will be able to gain sufficient trust in the other. Also, in order to be absolutely secure, many security decisions will have to be made explicitly by the user, which is contrary to the ubiquitous computing goal of reducing human intervention. Many applications will also require the free exchange of privileged information such as location, local capabilities, and constraints. Applications could run in an automated fashion if free exchanges were allowed, but privacy constraints could force a more conservative approach. Various service discovery and access mechanisms could also result in inadvertent exposure of private content and resources, owing to careless design or a lax policy. Submitting to privacy demands could detract from the user experience by restricting the performance of tasks. Alternatively, if the system cannot reconcile privacy demands with the task requirements, user intervention may be required. Privacy, therefore, will often be at cross-purposes with usability.

This three-way tradeoff severely impacts and potentially restricts security and privacy choices in ubiquitous computing, where usability and performance are key. Most research efforts in wireless networking and ubiquitous computing have emphasized the usability aspect at the cost of security and privacy [7, 27]. Though this results in a richer set of applications and functionality, a retrofitted security solution usually employs fairly rigid policies which interfere with many of the features that make the system usable. The approach we take is to analyze ubicomp interactions as a whole, rather than on a per-application basis. In this paper we attempt to identify the unique security threats and privacy and access control issues that are posed by device mobility and mutual anonymity of interacting devices and networks. In Sect. 8.2 we outline the threats posed by insecure infrastructure and malicious entities, and observe how mobility impacts systems in a negative way. In Sect. 8.3 we describe currently used and proposed approaches for maintaining security and privacy. We classify device-based security solutions into three categories, each providing security at a different level; this helps us to better understand and analyze these solutions.

8.2 Challenges of Unplanned Interactions

In the traditional computing paradigm, devices operate in a few established environments. Ubicomp necessitates a break from this pattern. Traveling from well-known and presumably safe environments to unfamiliar and potentially hostile ones poses many security challenges in mobile and pervasive computing. Likewise, the computing elements embedded in the infrastructure will encounter new and possibly unsafe devices all the time. Though a certain amount of paranoia is both healthy and necessary, it should not prevent devices from running essential tasks for users. Both users and their devices must take precautions. Devices should be able to verify the authenticity of the networking infrastructure, and the machines with which they communicate. Additionally, they must be able to assess the security risks in carrying out such interactions. Similar caution must be exercised by infrastructural components when interacting with unknown mobile devices that have entered communication range. Even if the external environment does not pose a threat, it may hardly be friendly. In these circumstances, protecting the integrity of system resources and data, as well as maintaining a necessary amount of privacy, is difficult. Challenges arise primarily due to communication with strangers, but in the absence of a trustworthy networking infrastructure, similar problems may afflict communication with known entities too. We address security and privacy issues both from an infrastructural and a device point of view; these issues include device and service provider authentication, the risks of habitual mobility, intelligent failure modes, and software agents. Challenges in each area must be addressed by researchers in order to achieve a complete security solution.

8.2.1 Infrastructure security and privacy

With traditional 802.3 Ethernet-based networking, when one plugs a device into a wall jack, it is typically assumed that the device receives connectivity from the local infrastructure. Clearly, there are possible attacks in this space, but in general this is a reasonable assumption since a physical wire acts as a physical metaphor tying the device to the physical environment. Wireless communications lacks this metaphor; absent policy, our mobile wireless devices can and will receive connectivity from any accessible service providers. This poses potential problems in that traditionally we have trusted our infrastructure to provide network services such as routing and name lookup. Malicious service providers can capture wireless clients and reroute requests to malicious services; such services are intended to duplicate legitimate services and capture personal identification information such as logins, passwords, credit card information, and so on. This type of session hijacking can be performed at the routing layer or by subverting DNS (domain name system).

There are several security problems here—one is the assumption that the networking infrastructure should provide routing and naming services in a secure and trusted manner; another is that one's device will associate with a given infrastructural component. These problems are related, especially if we seek to use trust relationships to deal with the former. The latter challenge is a problem of device authentication—i.e., how do we make sure we connect to the café's access point and not the malicious access point in a patron's backpack? This is a subset of the general device authentication problem—how do two mutually unknown devices authenticate one another?

Apart from ensuring the authenticity of the service provider whose network a mobile device is using, we must also deal with issues of data confidentiality and location privacy. These problems are exacerbated by the broadcast nature of the wireless medium, where eavesdropping is trivial for any device with a wireless card. Data confidentiality can be handled through encryption, and much research has gone into developing standards for 802.11 networks, which are mentioned in Sect. 8.3.1. But even if the communicated data cannot be interpreted, an eavesdropper can still infer the location of the communicating device and the entities it is talking to, which is information mobile users might want to keep private.

8.2.2 Device security and privacy

A number of security and access control problems lie within devices (or the end points of network connections) themselves. The problems arise due to misconfiguration, ineffective or bad security policies, vulnerable applications and insecure processes for remote discovery, access, and use of resources. Similar problems occur even in static desktop-based computing when communicating over the web, but the nature of devices in pervasive computing, mobility, and the frequency of contact with strangers worsens existing problems, as described below.

8.2.2.1 The Risks of Mobility

Mobility tends to exacerbate existing security and privacy challenges, such as system vulnerabilities and information leaks in network protocols. A mobile device moves in and out of environments with many unknown and potentially hostile devices, without the protection of infrastructure-based firewalls. This behavior exposes the device to more potential attackers, magnifying the risk of software vulnerabilities. When the mobile device is eventually taken home or to work, it passes behind traditional firewalls, possibly carrying an infection or an intruder.

A next-generation security system needs to be aware of these peripatetic devices that operate within its purview. The knowledge that a device is mobile and transient may allow the infrastructure to provide better support. Steps need to be taken to ensure the integrity of mobile devices and protect the rest of the local network from potential abuse. Challenges here include developing techniques to protect the network from mobile nodes while not overly inhibiting functionality.

8.2.2.2 Intelligent Failure Modes for Pervasive Security

Failure is an unfortunate fact of life. Mobile devices will be compromised, either over the network or by theft. It is incredibly important that the failure modes of such devices be engineered to minimize the impact of compromise. To that end, we need to focus on theft mitigation, reducing the ability to use or harvest data from a stolen device, as well as application limitations that restrict the powers of a compromised application, thereby protecting system integrity.

Theft Mitigation—Expensive and highly-portable mobile devices present tempting targets to thieves. In a time when identification theft is becoming all too common, these devices also represent a treasure trove of personal information. An important challenge thus is to mitigate the impact of theft—that is, reduce the utility of a stolen device, both in terms of actual functionality and in terms of extractable information. Additionally, recovery mechanisms including "phone home" features and secure remote localization capabilities would be valuable in the mobile device feature set.

Restricting Capabilities and Information Leaks—Mobility-oriented applications must be designed to limit the impact of compromise through segregation of functionality and by adopting the *least privilege* paradigm, limiting the application's privileges and data to those necessary to accomplish its tasks. This helps reduce the impact of malicious or compromised applications. Applications may deal with sensitive user data, including authentication information and financial data, as well as sensitive user context such as location or social relationships. A related challenge here is to limit the exposure of this data to the minimum necessary. Context can be made accessible at multiple fidelity levels, and only the necessary level of context should be exposed to the application. For example, location context can have levels such as "UCLA," "Boelter Hall," and "3564 Boelter Hall." The level of context exported to the application may depend on user policy, application needs, or the security characteristics of the local environment.

Similarly, the least privilege paradigm must be applied to information that is being transmitted. Remote computers should not be allowed to see more than is necessary for immediate purposes. Otherwise, information such as system or user identification information, system behavior patterns, etc., may be leaked to potentially hostile users. This information could be used by thieves to better target victims—i.e., the thief knows that one bus passenger has an expensive laptop and can determine which passenger, without even seeing the laptop. Similarly, if the presence of a given laptop in one's home is highly correlated with user presence, then radio emissions can be used to determine when someone is at home. In general, we need to be more careful about the radio emissions of our devices, as they do leak substantial information.

8.2.2.3 Software Agents and Mobile Code

Software agents and mobile code are frequently used in ubiquitous computing contexts to enable interoperability, application segmentation and migration, as well as customized handling of system operation. This raises serious security challenges. Mobile code may potentially harm the hosting device, or behave in unpredictable ways. The issuer of the mobile agent wishes to trust the result of the mobile code's execution, but the hosting device has control over the code. This poses a problem. Although this problem exists in the wired Internet, future pervasive environments may depend hugely on mobile agents to perform tasks, including the discovery of networks and services when devices are mobile. Such agents will be especially valuable in handling unplanned interactions.

Today's users already run a great deal of mobile code in the form of Java, JavaScript, Shockwave/Flash, and ActiveX controls. In many cases, mobile code intentionally or unintentionally has access to sensitive user data, often much more data than it strictly requires. We need reliable methods for protecting user data from disclosure and tampering while still permitting the execution of mobile code that is beneficial to the user. Accepting and running mobile code will require enhanced approaches for verification of code properties and establishment of trust.

8.3 Approaches

The concerns raised in the previous section can be summarized as: 1) protecting the integrity of the devices and networks, 2) preventing unnecessary data exposure, and 3) granting unknown entities permission to access private resources. As discussed in Sect. 8.1, enabling open interactions among mobile and infrastructure-based devices is a primary ubicomp goal. An impenetrable security system, though desirable in principle, would restrict access to many types of ubiquitous computing services. Instead, an effective system must be flexible in its approach to ensure both security and usability.

We can and must try to secure the networking infrastructure from malicious entities and eavesdroppers. Approaches to address this are discussed in Sect. 8.3.1. These will not solve the complete problem; traditional end-to-end security is still necessary. For the purposes of this discussion, we have chosen to define three subclasses within the solution space. While these subclasses are not exhaustive, we believe these are areas where further research could substantially address security and privacy challenges faced by most ubicomp scenarios.

- The *first* class of approaches (Sect. 8.3.2: *Resource/Content Protection and Access Control*) attempts to secure resources and content directly at the time of access. Such approaches also include situations where the device in question falls under the control of external entities, directly through theft or indirectly using mobile code.
- The *second* class of approaches (Sect. 8.3.2: *Secure Interaction Protocols*) comprises secure processes and protocols for interactions between devices, resulting in discovery of external resources and assignment of permissions to access those resources. The security and privacy solutions are managed by the device and are not tied to individual resources; the devices here are containers and controllers for a set of resources and services.
- The *third* class of approaches (Sect. 8.3.2: *Cross-Domain Security Frameworks*) consists of cross-domain security frameworks that impose security solutions in a top-down manner. Any two entities that come across each other in a pervasive computing world can determine the nature of their relationship and the scope of their interactions through such a shared framework. All trust frameworks, certificate hierarchies, and access control solutions for open systems fall under this category.

From one perspective, these three classes of solutions could form three layers of defense for any kind of interaction that takes place in a ubiquitous environment [14]. The trust approaches could help to determine the security basis for interaction among computing entities. Protocols could be used by such entities to discover each other's resources, securely configure permissions for access, and perform security-sensitive actions. At the innermost layer, once devices get to know each other's resource capabilities, they could directly access those resources which are guarded by low-level protection mechanisms. These three sets of approaches are neither mutually exclusive nor exhaustive. Furthermore, it is unlikely that a complete security solution can be drawn from any one of them alone. Trust frameworks are usually coupled with secure protocols for determining trust in external entities before permitting discovery and access. Resource protection mechanisms can be used in a scalable way in this context only if they are accompanied by a dynamic process of discovery and reconfiguration of local security state. An ideal security solution would combine appropriate features from all three classes of approaches that prove well suited to deployment in dynamic environments. Before we look at examples of different approaches from each of the categories defined above, we consider some mechanisms for securing network infrastructure.

8.3.1 Networking infrastructure security and privacy approaches

The most obvious technique used to maintain data confidentiality over any network link is encryption. As mentioned in Sect. 8.2, the broadcast nature of wireless communication makes this problem harder. Despite this, cryptographers and security engineers have developed workable security solutions for data confidentiality at the wireless MAC layer. Given the initial failure of the 802.11 WEP (Wired Equivalent Privacy) standard, [6], WPA (Wi-Fi Protected Access) was developed to overcome WEP's problems with stronger authentication schemes and a key management system. At higher layers in the network stack, devices have even more choices, and we can select from a variety of cryptographic schemes and key exchange protocols.

Preventing an eavesdropper from inferring the location of a device and the identity of the devices it is communicating with is still hard, mainly because of the broadcast nature of the communication medium. Also of interest is research in secure network discovery and connection to authentic service providers. This handles simultaneous discovery and authentication of a wireless network through automated means, which is complementary to the problem of private communication after connection establishment. Secure enrollment of a device to a network promises to mitigate the security problems associated with service provider selection and authentication, as described in Sect. 8.2.1.

8.3.1.1 Device Enrollment

The general problem of secure network enrollment within pervasive computing environments has been considered by several other projects. The canonical reference is Stajano and Anderson's *Resurrecting Duckling* [31] where the authors presented a model for imprinting wireless devices with network membership information through brief physical contact. In the model, physical contact is required to create a logical connection between two otherwise wireless devices. The *mother duck* controlling device would maintain absolute control over a set of duckling devices and their respective policies.

The duckling model has been further extended by the Palo Alto Research Center, Inc. (PARC) [1] and applied to home and enterprise-wide wireless LAN setup [2]. PARC removes the requirement for a secure side-band channel through the use of public key cryptography—this increases the baseline requirements for member devices, but allows more open side-band channels such as infrared. Recently, other approaches have investigated the use of embedded cameras to capture visual authentication information embedded in barcodes attached to devices [24], as well as the use of audio cues [18] coupled with displayed textual information.

The Instant Matchmaker [30] enables seamless and secure interactions between locally available devices and remotely situated personal resources through a form of short-lived enrollment. The Matchmaker functionality can reside on personal devices of limited capability that users carry around, such as a cell phone. This allows users to securely access their personal resources wherever they go, without having to carry them on their devices. A three-way enrollment protocol, using public keys, among the Matchmaker device, the local target device and the remote resource selected by the user, results in a time-limited secure connection between the target device and the remote resource.

8.3.2 Device-based security and privacy approaches

In this section we discuss approaches for maintaining security and privacy that are executed locally on devices. In general, these solutions assume the presence of a trusted communication infrastructure, though some trust-based solutions circumvent the networking problem altogether by enforcing stringent authentication schemes at the end points.

8.3.2.1 Resource/Content Protection and Access Control

In the world of pervasive and ubiquitous computing, data is often at risk for disclosure or tampering. Data lives on mobile and portable devices and may be subject to theft. One approach to protecting the privacy of user data is to integrate the protection mechanisms with the resources themselves.

Secure File Systems—Cryptographically secure file systems have been available for more than ten years [3, 36]. In practice, though, such file systems are not widely in use. Furthermore, even when such systems are used, it is common for users to store sensitive key material on the same device that is being protected. As a result, when devices are lost or stolen, it is likely that the information on those devices can be easily accessed by even modestly skilled attackers.

Additionally, when a device is taken over by malicious code, that code normally has full access to data on the device, including any encrypted data that the user may access. Typically, users rely on one master key or password to access their encrypted file systems. Thus, if the user accesses any encrypted data item, it is likely that all encrypted data items within that data-store are exposed to any malicious code that may be running on the device.

In order to protect data in this scenario, portable devices should not be the custodians of the key(s) to the sensitive data they hold. Rather, keys should be stored elsewhere and provided to applications on demand, based upon context and policy. If this were the case, certain data would be completely inaccessible to even the most determined attacker if the device was lost or stolen. Even in the case of device infection, much, if not all, sensitive data would be protected, ideally until the malicious code was discovered and purged.

Zero-Interaction Authentication—One system that possesses many of the properties mentioned above is Zero-Interaction Authentication (ZIA) [10]. In ZIA, each file is encrypted under a symmetric key, and that key is then encrypted with a key-encrypting key. A small security token, separate from the device itself, is the only entity that can decrypt file keys. The device must be in the presence of the token in order to access its own encrypted files. Thus, in our loss or theft scenario, ZIA cryptographically protects user data from disclosure from even the most determined adversary.

In addition to ZIA, other novel uses of cryptographic file systems and key management could greatly reduce the risk of disclosure of sensitive data through device loss or theft, or even device infection. Such systems should be informed by context and policy to provide more fine-grained and flexible control over encrypted data and associated keys than is currently provided by ZIA and other encrypted file systems.

Proof-Carrying Code—Although we can mitigate the dangers of device loss and theft, and we can to some extent limit the amount of sensitive data that is exposed in any particular context, it may be desirable or useful to run foreign code in various ubiquitous computing scenarios. Though many mobile code systems employ some facility for sand-boxing, much mobile code still has far more access than necessary, and often far more access than is safe. One possible approach to alleviating this problem is to use proof-carrying code [25]. In the ubiquitous world, devices will likely be offered mobile code from a variety of trusted and untrusted parties. In many cases, the user will explicitly run such code. In other instances, the device will be asked to run the code on behalf of the user. Proof-carrying code would maintain the usability we want, while preserving the safety and security of sensitive resources.

Proof-carrying code can provide proof of programmatic side-effects and invariants that can be reconciled with local policy. Depending on the level of trust (if any) ascribed to the provider of the code, the device can make safe and informed decisions without having to involve the user every time the question of executing mobile code is raised. Not only can proof-carrying code protect against malicious code that steals or tampers with sensitive user data, it can also preserve the overall integrity of the device, and may also have the added benefit of increasing the reliability of the device as a whole.

Proof-carrying code has addressed a very important problem, but we feel its complete potential has yet to be explored. A large number of ubicomp applications will depend on mobile code, and quick verification of security policy compliance would be very valuable. Application of proof-carrying code to ubicomp warrants further research.

8.3.2.2 Secure Interaction Protocols

Various situations will occur in ubiquitous computing where devices will need to discover each other's services and establish access permissions. The processes and protocols for managing secure discovery and assignment of access permissions comprise a different set of approaches, complementary to the resource protection mechanisms described above.

Trust Management—Trust management is a process that unifies security policies, credentials, authorization, and access control. This concept was introduced in PolicyMaker [4] and refined in KeyNote [5]. The process involves a request to perform a security-impacting action or to access private information or resources. The requestee runs a compliance checker taking as input the request, associated credentials from the requestor, and its local policies. If no conflict is detected, the request is granted; otherwise it is refused. This security or trust management solution requires a common trust framework, including a credential vocabulary, in order to be effective. In the mobile computing context, this solution maintains security and access control to the degree specified by the policies. One drawback is that the policies are static and are not sensitive to context changes. Although this process maintains the privacy and security of the requestee, it is not sensitive to the privacy considerations of the requester, who must provide all information and credentials demanded if the interaction is to succeed. Though both PolicyMaker and KeyNote were designed with traditional computing in mind, the technique could as well be used in pervasive computing when combined with a suitable process for discovery of networks and services.

Quarantine and Examination for Mobile Computing—We have explored a new paradigm for mobile and ubiquitous security called QED [15], or Quarantine, Examination, and Decontamination. In this paradigm, before mobile devices are allowed to join a wireless network, they are inserted into a quarantine zone. This is done to protect other local network participants from potential malware carried by the mobile device. While in quarantine, the device is subjected to an examination process that can include a variety of techniques such as external port scans and service identification, as well as internal tests that require cooperation of the device, such as virus scans and service patch determination. If problems such as vulnerabilities, undesirable services, or compromised software are found, the device may go through a decontamination phase in which the problems are, if possible, rectified. Once the infrastructure is confident that the device poses no threat, it is allowed to fully participate in the local network.

A system like QED demonstrates how security and privacy requirements may be at odds in a pervasive computing scenario. Security is enhanced if mobile devices run foreign code as instructed and report results truthfully. But this results in a loss of privacy for the device. Also, running arbitrary code itself requires a high measure of trust in the code provider. These are extremely important issues that require further research. The use of proof-carrying code techniques to verify policy compliance of examination modules deserves serious investigation. Also, verification of authenticity of returned examination results is an interesting problem; this could also have implications for digital rights management.

The Cisco Network Admission Control (NAC) system [9], a commercial product that is part of the Cisco Self-Defending Network Initiative, enforces access control in a domain through quarantine and examination. Access control decisions are based on a domain's security policies and involve checking incoming devices for vulnerabilities and infections. NAC suffers from certain drawbacks compared to QED; notably, it does not provide support for decontamination. Also, QED is completely software-based and open source, whereas NAC is integrated with Cisco hardware products. Using QED, security policies could be enforced in a flexible manner with access limits varying with degree of compliance. Also, the relationship between the mobile device and the network is more symmetric; this allows both the network and the mobile device to consider the privacy implications of running foreign code or releasing sensitive information. The primary goal of NAC is to enable domains to enforce security policies, and the relationship is inherently asymmetric. This solution will only work when a device interacts with familiar networks, and it is not flexible or scalable enough for ubicomp interactions.

Solutions performing QED functions are very valuable to mobile users who would be more tolerant of the added overhead. In the ubiquitous computing vision, applications must run smoothly in the face of frequent context changes. Scaling QED to work in those types of environments is well worth exploration.

Automated Peer Negotiation—We are exploring automated and flexible negotiation techniques among peers to enable interoperation among heterogeneous devices with diverse security and privacy policies [14]. Services can be discovered and resource access agreements can be reached via negotiation, while maintaining local security and privacy policies. Negotiation itself is not a new security mechanism, but rather ensures as much security as can be obtained through existing enforcement mechanisms. The policies, which are private to a system, describe the various constraints and inter-dependencies among system objects, and also describe the state of the system and the properties of its resources and mechanisms. The high level constructs are described in a common semantic language; we are leveraging Semantic Web frameworks like RDF (Resource Description Framework) and XML (Extensible Markup Language) for this purpose.

Negotiation is a flexible way for two entities in a ubicomp context to access each other's resources up to the maximum allowable risk and within the resource usage policies local to each. Most other approaches usually fall under extremes. At one end of the spectrum, some approaches for interaction obey rigid protocol semantics and are usually not applicable outside a particular domain. At the other end, open environments allow free and easy access without regard to security, such as early versions of Jini [32]. Negotiation offers a way to balance the risk of resource access or exposure of private information and the utility of permitting that operation. The crucial aspects are: 1) a trust/risk model that allows assessment of the risk associated with an operation or the trust gained in the other party, 2) a utility model that allows assessment of the benefits of gaining certain resources, and 3) a set of heuristic functions that allows an entity to determine when utility outweighs risk. Of course, there will be situations where the other party could be determined to be malicious, or mobile code found to contain a virus, in which case utility will rarely balance risk. The functions can be computed using the policies local to a system, which include user preferences as well as knowledge of security properties; e.g., risk of opening up a network port, how much trust does possession of certificate 'X' inspire, and so on. The negotiation protocol proceeds through a strategy whereby the parties can trade information, propose alternatives, and compromise within the limits of their policy constraints and the derived heuristic values. The policy language itself is backed by logical semantics and has a reasoning engine that enables query processing, knowledge chaining, and determination of conflicts. This is promising research, both from the security and privacy viewpoint and from the viewpoint of matching heterogeneous systems with available resources in a context-sensitive manner.

Negotiation as described above enhances the scope of prior work in automated trust negotiation [34], best illustrated by the TrustBuilder [35] and PeerTrust [17, 26] projects. Automated trust negotiation is a way of controlling access to a private resource over the web through a gradual process of trust building. In a typical instance of the protocol, requests for resource access generate counter-requests for credentials or other information, which in turn generate similar counter-requests. The process continues until a point of trust is reached or until failure occurs due to a conflict of privacy policies. Though trust negotiation was designed for the web, it can be adapted to the mobile and wireless context, though it would have to be augmented with secure discovery protocols. Through this process, resource access can be requested and obtained with minimum privacy loss for either party.

Zhu et al. [38] outline a service discovery protocol for pervasive computing which preserves privacy without third party mediation. The service provider and client expose partial sensitive information in a progressive approach. The protocol terminates when both parties reach an agreement about the extent of exposure of the service and authentication information. Upon a mismatch or an unsatisfied request, the protocol can be terminated without loss of privacy. This protocol is meant to handle fake service providers as well as unauthorized clients. Since entities are assumed to share low-level security information, which is the basis on which they negotiate, the scalability of this approach is debatable. Still, protocols of this type provide novel ways to maintain security and access control constraints in a decentralized manner without sacrificing openness.

8.3.2.3 Cross-Domain Security Frameworks

In a utopian world, all devices, networks, and enterprise domains would be completely open to any other entity that wished to interact with them. This is not practical, since every device cannot and does not trust every other device in mobile environments. Certain device properties, such as identity and relationships, reflect the amount of confidence that different humans have in each other, and by implication, affect device interactions. With perfect trust in the other party and in the communication channel, the process of interaction and the mechanisms used for resource and data access cease to matter. In practice, perfect trust is not feasible, especially when interacting entities are mutually anonymous. For example, a user could take his laptop to his office and immediately obtain access to the local network, as well as a range of other resources, given his role as a trusted member of that organization. Apart from basic authentication mechanisms that allow his laptop to connect and be admitted to the network, and similar authentication by the laptop to verify the network access point, strict security is generally not required for discovering the available resources or accessing privileged information. If the authentication framework and the process for handing out authentication information are foolproof, this will work. If a device is compromised or the owner turns malicious, there are serious consequences. If we put aside the issue of trusted entities turned malicious, having an overarching trust framework could enable free interoperation among any set of devices and networks. Such trust-based security solutions are commonly in use within limited domains, but an enterprise-based framework does not scale globally, and bottom-up growth of infrastructure also poses an obstacle to deployment. Below, we examine solutions that help in assessment of trust and discuss their advantages and drawbacks.

Centralized, Monolithic Security—A globally centralized security solution is a potential approach. Currently, efforts are being made to deploy single-provider, city-wide 802.11 network connectivity in a variety of metropolitan areas [23]. In theory, access to these services could be dependent on accepting a universal security policy. Every mobile device and network would be confident that all other entities would be constrained by that policy. This is conceptually a legitimate approach if it can be achieved at a worldwide scale, except for the fact that it would be undesirable to invest so much trust and power in one organization. This model creates a single point of failure which threatens user privacy as well as system reliability.

In the absence of a global security framework and policy, as well as an enforcement scheme, we need to devise frameworks for the dynamic establishment and assessment of trust in order to verify communication channels and enroll securely into foreign environments. These approaches are discussed below.

Certificate Hierarchies—The traditional distributed computing trust solution involves certificates. A certificate, in its simplest form, is a public key signed by certificate authorities. Gaining or verifying trust using certificates requires a hierarchy of certificate authorities. An ad hoc interaction could involve the presentation of a certificate; if the recipient shares a common parent with the certificate owner at some level in the hierarchy, a trust relationship can be established. Though this approach provides a certain degree of trust in mobile and ubiquitous computing, it has serious drawbacks which limit its use. First, given the bottom-up growth of ubicomp infrastructure, it is difficult to force everyone to accept one particular certificate hierarchy, and the higher up the common authority lies, the lower the value of trust becomes. Second, with a huge and unwieldy infrastructure, revocation and updates will be very inefficient. Third, this does not handle cases where strangers meet in a virtual bubble, possibly having no connection with a common trust authority. Last, and most important, certificates in their basic forms (or the way they are currently used in web transactions) are identity-based, and do not say anything more; every mobile device or network has different concerns and priorities, and simply verifying that a particular authority has certified the opposite party may not mean anything.

Peer-to-Peer Trust—Delegation has been proposed and used by various researchers to make the certificate distribution and verification scheme less strictly hierarchical and more suited to dynamic mobile environments. For example: entity A could delegate to entity B the right to issue certificates in A's name. Therefore, a delegated certificate issued by B could be trusted if A is a trusted source. This scheme has the property of creating chains and webs of trust [39], which effectively form a peer-to-peer security framework that could be used as a basis for interaction. Though more dynamic, decentralized, and more resilient to network partitions, this kind of framework suffers from the same problems that afflict certificate hierarchies; it is difficult to assess the value of a credential issued by any particular peer. What makes the issuer of the credential trust a particular entity is not clear, especially if the distance along the chain between the certificate owner and the examiner is long. Clearly these delegated credentials need to provide more information than just identities. In this respect, we are building a voucher mechanism in which a voucher can be provided by one entity to another, certifying certain properties such as rights, group affiliation, and state. The use of a rights-delegating voucher is similar to SPKI (Simple public key infrastructure) [11].

Closely associated with webs and chains of trust is the notion of reputation, which in theory adds some more weight to the trust or confidence level in another party. Reputation is a way of assessing the trustworthiness of entities based on what other known and trusted entities say about them [37]. If this were to work, it would be a strictly more reliable framework than one based on identity. Reputation models have not seen much success due to the impact of lying or colluding parties, and the huge number of variables involved in trust assessment [28]. Still, this is one way of establishing an overarching web of trust that could potentially cover most unplanned ubicomp interactions, and research in this area should be watched closely.

Role-based access control (RBAC) is a popular security framework adopted by open systems, where privileges are tied to a defined role. In its simplest form, this kind of access control works in the mobile context only if familiar entities interact. If strangers must interact securely, the system must be augmented by some process of role determination. Given a common credential vocabulary, a web of trust, and delegation permissions, privileges can be determined through a recursive process of proof-building, as demonstrated in the dynamic RBAC model [16]. Combining role-based access control with delegation and trust chains has been employed in ubicomp middleware like Centaurus [20] and Vigil [21, 22].

Quantitative Trust Models—Newer approaches have argued for a more dynamic notion of trust, and one that reproduces the way humans interact among themselves, such as the Secure project [8, 12]. The dynamic nature of trust can be reproduced through the processes of trust formation and trust evolution, both of which use the history of past interactions in the trust evaluation functions. This project, as its basis, advocates making personal observations of an entity's behavior a part of the trust assessment function. A system for monitoring applications and reacting to events [13] is based on such dynamic trust models. This is a promising approach for managing dynamic environments, as it has the best potential for allowing secure interactions among strangers. Apart from identifying the important features of a trust framework, we need quantitative models to generate and make use of trust relationships. One approach could be a unified model that uses both identity and contextual properties and which expresses trust as a continuum [29]. A different model attempts to model trust using probabilities, and in addition proposes ways to interpret the information during the actual process of performing a security-sensitive action [19].

We feel that dynamic trust models of the type discussed above hold great promise, and indeed are some of the few trust frameworks that scale to ubicomp environments. We cannot of course abandon identity and possession of certificates as a means of assessing trust; these are and will be key mechanisms for trust building. Therefore, research must concentrate on producing trust frameworks that make use of identity, properties, and observed results of actions. These kinds of trust frameworks also form the basis of automated peer negotiation, which was discussed earlier, and this is a promising research area that we are actively investigating.

8.4 Conclusion

We have discussed a wide spectrum of security and privacy issues that must be addressed before we can trust our devices to perform automated tasks on our behalf in a mobile context. Trustworthy and secure communication infrastructure is a prerequisite for secure mobile computing. Our own mobile devices and the other devices they interact with in the environment must have security and privacy solutions built in so that they can discover and access each other's resources even when connections are established in an ad hoc manner. In a ubiquitous computing world, usability is of primary importance, and security and privacy solutions must be designed in such a way that they preserve this property.

We have classified device-based solutions into three categories, roughly corresponding to three layers of defense for a mobile or infrastructure-based device interacting in dynamic circumstances with entities that may or may not be familiar. Each class of solutions has drawbacks if employed in isolation. Resource or content protection mechanisms employed without secure protocols for discovery and a trust basis either provides weak security (for interactions with strangers) or does not scale and would require some amount of manual configuration. Similarly, a secure negotiation protocol for sharing of resources without the enforcement mechanisms at the resource access level or a trust basis is not a comprehensive security solution. Trust frameworks without secure means of trust inference and enforcement at lower levels do not provide much value. A hybrid of the three classes of approaches is required for a scalable security solution, and for mobile devices to trust their surrounding environment and service providers when interactions are required.

We have also identified a number of promising approaches that address security and privacy challenges faced by mutually unknown entities interacting in an unplanned manner. We envision secure enrollment schemes growing in importance. More applications inevitably lead to more software vulnerabilities, and QED-like integrity analysis will be indispensable for halting the spread of malware. Some flavor of negotiation will inevitably come into play when interacting with strangers, since this promises to address the subtle balance required between security, privacy, and usability. Trust frameworks that are not purely identity-based are the weak point in today's research, and further investigation in this area would be very welcome.

We can assume that decentralized operation and numerous unplanned interactions will be predominant features of emerging ubiquitous computing systems. Dealing with unknown entities and unplanned events will pose numerous challenges. By limiting the risks of exposure and compromise at multiple levels, systems may remain secure, despite the dangerous and hostile intent of others. Taking lessons from the approaches discussed in this paper, future security framework designs must focus on risk minimization as a primary goal.

References

1. Balfanz D, Smetters DK, Stewart P, Wong HC (2002) Talking to Strangers: Authentication in Ad-Hoc Wireless Networks. In: Proceedings of the 9th Network and Distributed System Security Symposium, San Diego, California, The Internet Society, pp 23–25

2. Balfanz D, Durfee G, Grinter R, Smetters DK, Stewart P (2004) Network-in-a-Box: How to Set Up a Secure Wireless Network in Under a Minute. In: Proceedings of the 13th Usenix Security Symposium, San Diego, California, pp 207–222
3. Blaze M (1993) A cryptographic file system for UNIX. In: Proceedings of the 1st ACM Conference on Computer and Communications Security, ACM Press, New York, New York, pp 9–16
4. Blaze M, Feigenbaum J, Strauss M (1998) Compliance Checking in the PolicyMaker Trust Management System. In: Proceedings of the Financial Cryptography Conference, Lecture Notes in Computer Science, vol 1465. Springer-Verlag, London, UK, pp 254–274
5. Blaze M, Feigenbaum J, Ioannidis J, Keromytis AD (1999) RFC 2704 - The KeyNote Trust Management System Version 2. RFC 2704, Network Working Group
6. Borisov N, Goldberg I, Wagner D (2001) Intercepting Mobile Communications: the Insecurity of 802.11. In: Proceedings of the 7th annual International Conference on Mobile computing and networking, Rome, Italy, pp 180–189
7. Brooks R (1997) The Intelligent Room Project. In: Proceedings of the 2nd International Cognitive Technology Conference, Aizu, Japan
8. Cahill V, Gray E, Seigneur J, Jensen CD, Chen Y, Shand B, Dimmock N, Twigg A, Bacon J, English C, Wagealla W, Terzis S, Nixon P, di Marzo Serugendo G, Bryce C, Carbone M, Krukow K, Nielsen M (2003) Using Trust for Secure Collaboration in Uncertain Environments. IEEE Pervasive Computing Journal, vol 2, no. 3, pp 52–61
9. Cisco Systems (2003) Network Admission Control Executive Positioning Document. White Paper—Cisco Network Admission Control, http://www.cisco.com/en/US/netsol/ns466/networking_solutions_white_paper0900aecd80 0fdd66.shtml
10. Corner M, Noble B (2002) Zero-Interaction Authentication. In: Proceedings of the 8th annual International Conference on Mobile Computing and Networking (MobiCom), Atlanta, Georgia, pp 1–11
11. Ellison C, Frantz B, Lampson B, Rivest R, Thomas B, Ylonen T (1999) RFC 2693 - SPKI Certificate Theory. RFC 2693, Network Working Group
12. English C, Nixon P, Terzis S, McGettrick A, Lowe H (2002) Dynamic Trust Models for Ubiquitous Computing Environments. In: Proceedings of the First Workshop on Security in Ubiquitous Computing at the Fourth annual International Conference on Ubiquitous Computing, Göteborg, Sweden
13. English C, Terzis S, Nixon P (2004) Towards Self-Protecting Ubiquitous Systems: Monitoring Trust-based Interactions. Personal and Ubiquitous Computing Journal, vol 10, issue 1 (December 2005), Springer London, pp 50–54
14. Eustice K, Kleinrock L, Markstrum S, Popek G, Ramakrishna V, Reiher P (2003) Enabling Secure Ubiquitous Interactions. In: Proceedings of the 1st International Workshop on Middleware for Pervasive and Ad-Hoc Computing at the 4th ACM/IFIP/USENIX International Middleware Conference, Rio de Janeiro, Brazil
15. Eustice K, Kleinrock L, Markstrum S, Popek G, Ramakrishna V, Reiher P (2003) Securing WiFi Nomads: The Case for Quarantine, Examination, and

Decontamination. In: Hempelmann C, Raskin V (eds) Proceedings of the New Security Paradigms Workshop, Sponsored by Applied Computer Security Associates, Ascona, Switzerland, pp 123–128
16. Freudenthal E, Pesin T, Port L, Keenan E, Karamcheti V (2002) dRBAC: Distributed Role-Based Access Control for Dynamic Coalition Environments. In: Proceedings of the 22nd International Conference on Distributed Computing Systems, IEEE Computer Society, Vienna, Austria, pp 411–420
17. Gavriloaie R, Nejdl W, Olmedilla D, Seamons K, Winslett M (2004) No Registration Needed: How to Use Declarative Policies and Negotiation to Access Sensitive Resources on the Semantic Web. In: Proceedings of the 1st First European Semantic Web Symposium, Heraklion, Greece, Springer-Verlag, pp 342–356
18. Goodrich M, Sirivianos M, Solis J, Tsudik G, Uzun E (2005) Loud and Clear: Human-Verifiable Authentication Based on Audio. In: Proceedings of the 26th IEEE International Conference on Distributed Computing Systems, p 10
19. Jøsang A (1999) Trust-Based Decision Making for Electronic Transactions. In: Proceedings of the Fourth Nordic Workshop on Secure IT Systems, Stockholm, Sweden, Stockholm University Report, pp 99–105
20. Kagal L, Korolev V, Chen H, Joshi A, Finin T (2001) Centaurus: A Framework for Intelligent Services in a Mobile Environment. In: Proceedings of the International Workshop on Smart Appliances and Wearable Computing at the 21st International Conference on Distributed Computing Systems, Mesa, Arizona, pp 195–201
21. Kagal L, Finin T, Joshi A (2001) Trust-Based Security in Pervasive Computing Environments. IEEE Computer Journal, vol 34, no. 12 (December 2001), pp 154–157
22. Kagal L, Undercoffer J, Perich F, Joshi A, Finin T (2002) A Security Architecture Based on Trust Management for Pervasive Computing Systems. In: Proceedings of Grace Hopper Celebration of Women in Computing
23. Kopytoff V, Kim R (2005) Google offers S.F. Wi-Fi—for free / Company's bid is one of many in response to mayor's call for universal online access. Article: San Francisco Chronicle, http://www.sfgate.com/cgi-bin/article.cgi?file=/c/a/2005/10/01/MNGG9F16KG1.DTL, October 1, 2005
24. McCune JM, Perrig A, Reiter MK (2005) Seeing is Believing: Using Camera Phones for Human-Verifiable Authentication. In: Proceedings of the IEEE Symposium on Security and Privacy, Oakland, California, pp 110–124
25. Necula G (1997) Proof-Carrying Code. In: Proceedings of the 24th Annual ACM SIGPLAN-SIGACT Symposium on Principles of Programming Languages, Paris, France, pp 106–119
26. Nejdl W, Olmedilla D, Winslett M (2004) PeerTrust: Automated Trust Negotiation for Peers on the Semantic Web. In: Jonker W, Petkovic, M (eds) Proceedings of the VLDB 2004 International Workshop on Secure Data Management in a Connected World, Lecture Notes in Computer Science, vol 3178, pp 118–132
27. Román M, Hess C, Cerqueira R, Ranganathan A, Campbell R, Nahrstedt K (2002) A Middleware Infrastructure for Active Spaces. IEEE Pervasive Computing Journal, vol 1, issue 4 (October 2002), pp 74–83

28. Sen, S, Sajja N (2002) Robustness of Reputation-Based Trust: Boolean Case. In: Proceedings of the First International Joint Conference on Autonomous Agents and Multiagent Systems: part 1, Bologna, Italy, International Conference on Autonomous Agents, pp 288–293
29. Shankar N, Arbaugh WA (2002) On Trust for Ubiquitous Computing. In: Proceedings of the First Workshop on Security in Ubiquitous Computing at the Fourth annual International Conference on Ubiquitous Computing, Göteborg, Sweden
30. Smetters DK, Balfanz D, Durfee G, Smith TF, Lee KH (2006) Instant Matchmaking: Simple and Secure Integrated Ubiquitous Computing Environments. In: Dourish P, Friday A (eds) Proceedings of the 8th annual International Conference on Ubiquitous Computing, Orange County, California, Springer, pp 477–494
31. Stajano F, Anderson R (1999) The Resurrecting Duckling: Security Issues for Ad-hoc Wireless Networks. In: Proceedings of the 7th International Workshop on Security Protocols, Lecture Notes in Computer Science, vol 1796, Cambridge, UK, pp 172–194
32. Waldo J (1999) The Jini Architecture for Network-Centric Computing. Communications of the ACM Journal, vol 42, no. 7 (July 1999), pp 76–82
33. Weiser M (1991) The Computer for the 21st Century. Scientific American Magazine, vol 265, no. (September 1991), pp. 94–104
34. Winslett M (2003) An Introduction to Trust Negotiation. In: Proceedings of the 1st International Conference on Trust Management, Heraklion, Greece, Lecture Notes in Computer Science, vol 2692, pp 275–283
35. Winslett M, Yu T, Seamons KE, Hess A, Jacobson J, Jarvis R, Smith B, Yu L (2002) Negotiating Trust on the Web. IEEE Internet Computing Journal, vol 6, issue 6 (November 2002), pp 30–37
36. Wright CP, Martino M, Zadok E (2003) NCryptfs: A Secure and Convenient Cryptographic File System. In: Proceedings of the USENIX 2003 Annual Technical Conference, pp 197–210
37. Xiong L, Liu L (2004) PeerTrust: Supporting Reputation-Based Trust in Peer-to-Peer Electronic Communities. In: IEEE Transactions on Knowledge and Data Engineering, vol 16, no. 7 (July 2004), pp 843–857
38. Zhu F, Zhu W, Mutka MW, Ni LM (2005) Expose or Not? A Progressive Exposure Approach for Service Discovery in Pervasive Computing Environments. In: Proceedings of the Third IEEE International Conference on Pervasive Computing and Communications, Kauai Island, Hawaii, pp 225–234
39. Zimmermann PR (1995) The Official PGP User's Guide. The MIT Press, Cambridge, Massachusetts

9 An Anonymous MAC Protocol for Wireless Ad-hoc Networks

Shu Jiang

Department of Computer Science, Texas A&M University

9.1 Introduction

A wireless ad hoc network is formed by a set of mobile hosts that communicate over wireless medium such as radio. Due to ease of deployment, it has many mission-critical applications in military as well as in civilian environments. Those applications usually have strong requirements for data confidentiality and privacy. In this chapter, we address one of the most challenging confidentiality and privacy issue with wireless ad hoc networks: anonymity of communication.

Communication anonymity entails the hiding of information that two hosts communicate with each other. In general, there are three ways of achieving this goal, i.e., hiding the source, hiding the destination, or hiding the source-destination combination of a communication [10]. In wireless ad hoc networks, all communications over a network are vulnerable to eavesdropping. A connection between two hosts can be exposed by the source and destination fields in the headers of data packets sent over the connection. As a solution, the two hosts can set up an *anonymous connection* between each other and encrypt each data packet in such a way that the two hosts never appear in the source and the destination fields of the packet header simultaneously [12]. However, this solution requires that all data packets of a connection follow the same and predetermined routing path for delivery, while in a wireless ad hoc network, there is no guarantee of a fixed routing path between any two hosts, due to node mobility and changing topology. To overcome this problem, a set of anonymous routing protocols were proposed recently in the literature [7, 3, 14].

ANODR [7] is an anonymous on-demand routing protocol for wireless ad hoc network. It has two functions. First, it discovers a route between two hosts on demand and sets up an anonymous connection. Second, when the route is broken, it repairs the route or discovers a new route and maintains the connection. On a multi-hop route, each hop is assigned a unique *route pseudonym* and each intermediate node stores the mapping between the route pseudonyms of its previous hop and its next hop in a forwarding table. When a data packet is sent, both its source and destination addresses are masked, and it is forwarded based on the route pseudonym it carries. In the beginning, it carries the route pseudonym of the first hop on its route. The source host then broadcasts the packet within its transmission range. After receiving the packet, the first node will look up its forwarding table, modify the packet to carry the route pseudonym of the next hop, and broadcast the packet. So after each transmission, the packet will carry a different route pseudonym. In addition, each intermediate node also changes the appearance of the packet (i.e., bit pattern) and uses mixing techniques [4] such as random delay to thwart all tracing attempts.

From the above description, we see that ANODR utilizes the link-layer broadcast and link layer encryption mechanism during data forwarding process. In order to improve reliability of link layer broadcast, it uses a simple anonymous acknowledgment protocol. In the protocol, upon receipt of a data packet, the receiver node should locally broadcast an anonymous ACK packet. Obviously, there exists a timing link between a data packet and its triggered ACK packet, which can be utilized by an eavesdropper to deduce the intended receiver of a data packet. ANODR assumes that an eavesdropper can only learn the transmitting node of a packet from its MAC address and sets it to all-1's. Unfortunately, this is not a sound assumption. There are technologies for locating a transmitting node based on physical layer characteristics such as signal strength [1, 13]. In addition, the adversary can deploy many near-invisible sensors (e.g., camera) to locate and track all node movements in a particular area. In this situation, ANODR cannot meet its reliability requirements without compromising anonymity.

In this chapter, we propose a MAC protocol to address the needs for anonymity and reliability with respect to link-layer broadcasts simultaneously. Our protocol is resistant against powerful eavesdroppers we described above, who can reveal the senders of all transmissions. In our protocol, each node broadcasts a batch of data packets, instead of one data packet, at a time. The packets in the batch may be addressed to different receivers. It is possible that some packets are lost due to collisions or interferences. In order to deliver as many packets as possible, the sender needs to query every receiver about their receiving status and decide which packets need to be retransmitted. This is achieved by a polling scheme. The sender selects a subset of neighbors and sends POLL messages to each of them individually. Each node being polled should send a REPLY message back. All messages are encrypted, which contain information such as the sequence numbers of received packets. The polling list is constructed independently from the list of receivers to which data packets have been sent. So the adversary cannot build strong links between the two lists. The rest of the chapter is organized as follows. In Sect. 2, we describe the details of the protocol design. In Sect. 3, we present a security analysis of the protocol. In Sect. 4, we show the performance evaluation results of the protocol obtained from ns-2 [2] simulations. Sect. 5 is a summary of the chapter.

9.2 Protocol Design

In this section, we describe the details of the proposed anonymous MAC protocol. To conform with the IEEE 802.11 protocol [5], we call units of transmission as "frames", instead of packets. This protocol serves two purposes. First, it can hide the receiver of a unicast data frame. This is achieved by transforming a unicast frame to a broadcast frame and encrypting the receiver node address along with the frame payload. We assume that the sender and receiver share a secret WEP key. Since the receiver of a transmitted data frame is not identified by an explicit node address, each node within the sender's transmission range is possible. These nodes comprise the "anonymity set" [10] for the frame. Second, it provides reliability for anonymous data frames. This service is provided under the premise that it does not compromise receiver anonymity of the frames. We assume a strong adversary model, where the adversary can link the source of each transmission to a particular node. In other words, there is no source anonymity of frames. We design a sender-initiated polling mechanism to achieve the goal. In the following, we first define the formats of control frames and anonymous data frame, and then describe the sender's protocol and the receiver's protocol.

9.2.1 Frame format

Figure 9.1 shows the format of a POLL frame. The RA is the address of the node being polled, and the SA is the address of the node transmitting the POLL frame. The duration value is the time required to complete the current poll, which is calculated as the transmission time of a REPLY frame plus one *SIFS* interval. The IV is the initiation vector used in WEP encryption. The sequence number is explained below. The padding is a number of random bytes produced to prevent content attack (explained in Sect. 4). The last two fields comprise the plaintext for encryption.

Figure 9.2 shows the format of a REPLY frame. The RA is the address of the node transmitting POLL. The sequence number and bitmap fields are used by the ARQ protocol (explained below). The padding field has the same function as in POLL frame.

Figure 9.3 shows the format of an anonymous data frame. The pseudo header has three fields: RA is the address of the intended recipient node, Sequence is the sequence number assigned to the frame, Padding is a number of random bytes.

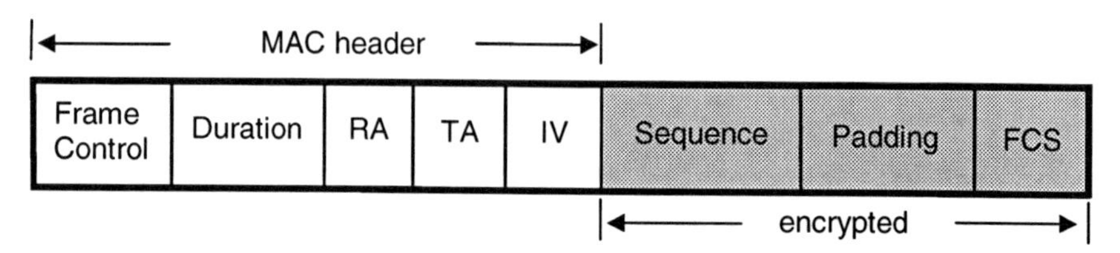

Fig. 9.1. POLL frame format

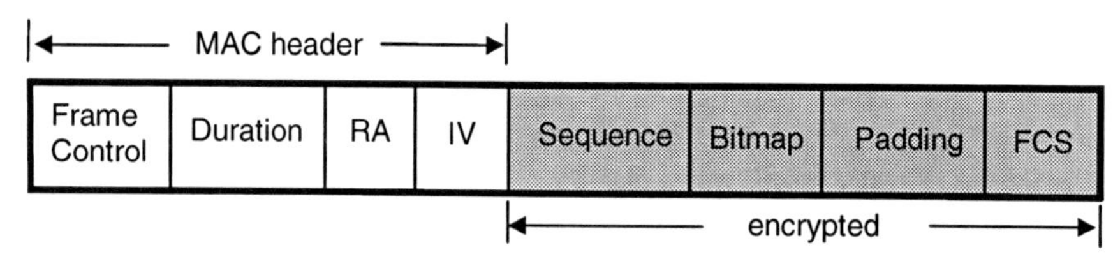

Fig. 9.2. REPLY frame format

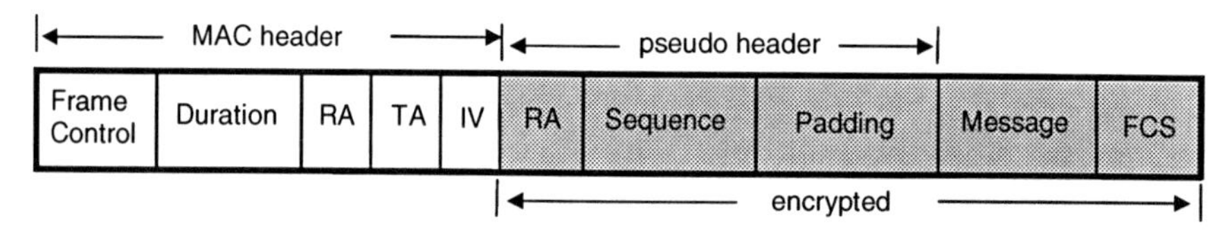

Fig. 9.3. Anonymous data frame

9.2.2 Sender's protocol

Each node maintains a FIFO queue, holding frames that are waiting to be transmitted or retransmitted. When a new frame is received from the upper layer, it is given a sequence number. The sender and receiver use this sequence number to track and retransmit lost frames. For this purpose, each node i maintains a variable SN_{ij} with respect to each neighbor node j. SN_{ij} is initiated to 0 at the system setup time. For each new frame transmitted to j, node i assigns SN_{ij} to the frame and increments SN_{ij} by 1. This ensures that node j receives frames from node i with contiguous sequence numbers. If a number is missing, the frame must be lost during transmission.

At each node i, with respect to each neighbor node j, a sending window $[LSN_{ij}, HSN_{ij}]$ is maintained to record the range of sequence numbers of frames stored in the queue. LSN_{ij} is the lowest sequence number of frames, from i to j, currently in the queue, while HSN_{ij} is the highest sequence number. Node i advances LSN_{ij} in two cases:

1. Node j acknowledges receiving of the frame with sequence number LSN_{ij};
2. Node i fails to transmit the frame with sequence number LSN_{ij} after a maximum number of attempts and discards it.

At each node i, if the queue is not empty, the following algorithm is executed:

1. Node i follows the CSMA/CA protocol in IEEE 802.11 to obtain the right to transmit. It works as follows. The node first senses the channel. If the channel is busy, it just waits until the channel becomes idle. If the channel has been idle for at least *DIFS* period (= $50\mu s$), the node enters a state of collision avoidance and backs off from transmitting for x slots of time, where x is a random number within the contention window. In the collision avoidance state, if the channel is sensed busy, the node will suspend its backoff timer immediately and resume the timer only after the channel is again sensed free for a *DIFS* period. When the backoff timer counts down to zero, go to step 2.
2. Node i constructs a polling set by adding all receivers of data frames currently in the queue. If the polling set size is smaller than a preset value *MIN_POLLING_SET_SIZE*, it randomly chooses nodes within the transmission range to add in.
3. Node i polls nodes in the polling set at a random order. If a polled node is j, the corresponding POLL frame has the current value of LSN_{ij} in its sequence field. For each polled node, after node i transmits the POLL frame, it switches to the receiving mode and waits for reply. If the channel is still free after two *SIFS* intervals, node i assumes that the polled node does not receive the POLL frame and starts polling the next node. If a valid REPLY frame is received from the polled node, node i will update its state based on the information in it (e.g., releasing acknowledged frames, advancing the sending window, incrementing retry counters of unacknowledged frames), and polls the next node after one *SIFS* interval. If node i receives a corrupted REPLY frame or senses a busy medium during the *SIFS* interval, it will follow the binary exponential backoff algorithm in 802.11 and go to step 1.

4. If all nodes in the polling set have been polled, the nodes from which REPLY frames are successfully received are "available receivers". Node i transmits only frames to available receivers in the queue. So some frames may be skipped. For a retransmitted frame, node i needs to change the padding value in the pseudo header and reencrypt the frame. Consecutive frames are spaced by *SIFS* intervals. There is a maximum number of frames that can be transmitted in a batch. This is a system parameter (referred as *MAX_BATCH_SIZE*) whose value affects the system performance. In our experiments, we set *MAX_BATCH_SIZE* to 4. The possibility exists, especially when network load is extremely high, that node i received no REPLY frames from any polled nodes. In this case, node i would abort the transmission, follow the binary exponential algorithm and go to step 1. If a node fails to reply consecutive pollings for a maximum number of times, the link is assumed to be broken and all frames to be sent on that link are purged from the sender's queue.

9.2.3 Receiver's protocol

At each node j, with respect to each neighbor node i, a receiving window is maintained to record the sequence numbers of received frames. In Selective Repeat ARQ protocol, a common approach is to use two variables to implement a receiving window: a Lowest Bound LB_{ji} and a one-byte Bitmap BM_{ji}. All frames from i with sequence numbers lower than LB_{ji} have been received. The BM_{ji} indicates the receiving status of frames whose sequence numbers higher than LB_{ji}. Specifically, if the k-th bit of BM_{ji} is 1, it means that the frame with sequence number $LB_{ji}+k$ has been received. For example, a LB_{ji} of 100 and a BM_{ji} of 11100110 indicate that node j has correctly received frames 0–99, 101, 102, 105, 106, 107, whereas frames 100, 103, 104 were lost. Node j advances its receiving window in two cases:

1. When a POLL from node i is received, if $LSN_{ij} > LB_{ji}$, it means that the sender node i has advanced its sending window and given up its attempts to retransmit frames lower than LSN_{ij}. This could happen when node j experienced temporary severe interference. In this case, node j synchronizes its receiving window with node i's sending window by advancing LB_{ji} to LSN_{ij}.
2. When a data frame from node i is received, if its sequence number matches with LB_{ji}, then node j can advance its receiving window, i.e., incrementing the LB_{ji} by 1 and right-shifting the BM_{ji} for one bit. Node j can repeat the adjustment until the lowest bit of BM_{ji} is 0. If the sequence number of the received data frame is larger than LB_{ji} and is not a duplicate, the BM_{ji} is updated to indicate the receiving status.

Unlike many Selective Repeat ARQ based protocols, we do not maintain a "receiver buffer" at the MAC layer to hold out-of-sequence frames. Instead, a receiver passes each received frame immediately to the upper layer (i.e., network). There are two reasons. First, this reduces the queueing delay. Second, frames transmitted on a link belong to different end-to-end flows and typically have different next hop receivers. Frame loss of one flow should not affect the frame delivery of other flows. This is similar to the head-of-line problem in router design. By relaxing the in-sequence constraint, we can increase the overall network throughput. Notice that to provide reliable message delivery for users, the destination node now has responsibility for sequencing.

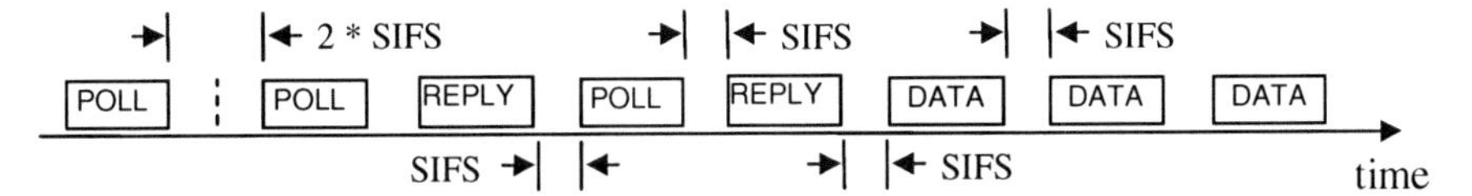

Fig. 9.4. An illustration of the scheme

The described protocol is illustrated in Fig. 9.4. In the figure, the first polled node does not send a REPLY frame, probably not receiving the POLL. Therefore, the sender sends the second POLL (to a different node) after two *SIFS* intervals. Since any node can transmit if the channel remains free for *DIFS*, having sender transmitting the second POLL earlier, without waiting for the transmission time of a REPLY frame, prevents any neighbor from interrupting the polling process. The second and third POLLs are replied. Each polled node transmits the REPLY frame immediately, after one *SIFS* interval. Data frames in the current batch are transmitted continuously, with one *SIFS* spacing between two consecutive frames. So, during the entire process, the medium is never idle for more than $2 \times SIFS$.

9.3 Security Analysis

In this section, we present a security analysis of the protocol. The objective of an adversary is to trace a packet from its source to its destination. To achieve this goal, the adversary needs to reveal the receiver of the packet at each hop while it is being forwarded. In our protocol, the receiver address at each hop is encrypted in the pseudo header of the packet. We assume that the adversary is not capable of breaking the link encryption through cryptanalysis. He or she has only two choices. One is to compromise nodes. Another is to launch traffic analysis attack.

9.3.1 Compromised node

If a node is compromised, the adversary can immediately reveal partial route of each packet forwarded by the node. Whether the entire route of a packet can be revealed depends on whether there are enough compromised nodes on the route such that the exposed segments can be linked together. Kong et al's analysis on route traceability in the presence of compromised nodes also applies here [7].

When there are compromised nodes in a sender's neighbor set, the maximum receiver anonymity that can be achieved for a packet is determined by the number of uncompromised nodes in the set. In the current design, the polling set is a subset of the sender's neighbor set. A more secure design is to make the polling set be exactly the sender's neighbor set. However, our simulation results show that the performance of this design would be very poor when the average node degree is more than 6. The current design tries to implement a trade-off between security and performance.

9.3.2 Traffic analysis attack

For a conventional MIX, the attacker tries to find correlation between an input message and an output message of the MIX. To achieve this goal, the attacker can utilize message content, size, timing information, or can manipulate the input and output messages. Specifically, *content attack* compares the contents of two messages bit by bit, looking for match; *size attack* examines the message lengths and is only effective against protocols using variable-length messages; *timing attack* searches for temporal dependencies between transmissions. *Flooding attack* (aka. *node flushing attack*, *n-1 attack*) is a special form of content attack. In case of a simple threshold n MIX, which flushes after receiving n messages, the attack proceeds as follows: When the attacker observes a targeted message entering the MIX, it sends n-1 messages into the MIX to make it fire. Since the attacker can recognize all his own messages when they leave the MIX, the remaining one must be the targeted message and its destination is revealed.

The above description of traffic analysis attacks applies to MIXes in a switching network. In an anonymous broadcast network, each attack may take a bit different form, in that the attacker searches for correlation between apparently independent transmissions by different nodes (see Fig. 9.5). For example, node A transmits a frame at time t, and node B, one of its neighbors, transmits at time $t + \varepsilon$. This may suggest that node B is the receiver of node A's frame and is forwarding the frame to its next hop.

However, for this timing attack to succeed, the following conditions must be satisfied:

1. The queue is empty when node B receives the frame, and
2. All other neighbors of node A have no frames to transmit.

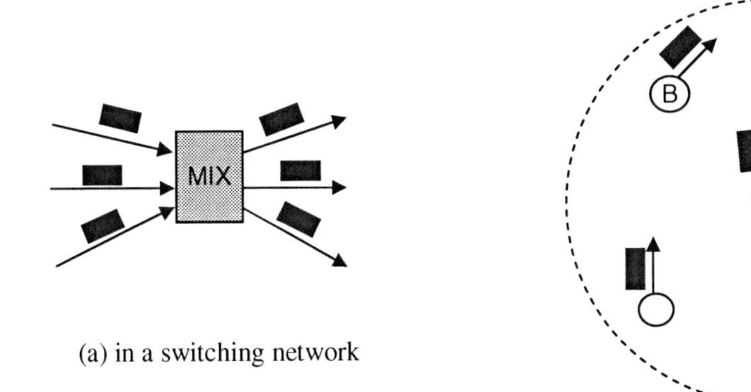

(a) in a switching network

(b) in an anonymous broadcast network

Fig. 9.5. Different attacking scenarios against MIX

If any of the above conditions is not satisfied, then the probability of a successful attack would be reduced, due to a larger delay between two transmissions of the same frame. This suggests that each node having a non-empty queue, i.e., always in saturation mode, has benefits to security. The queue here serves a similar function as the "pool" in a conventional MIX. Again, there is a trade-off between security and performance. In the current design, the scheme does not generate dummy data frames, and only generates dummy polls, based on the assumption that network users provide enough traffic loads. However, it can be easily extended to apply to low-traffic networks, by allowing nodes to generate dummy data frames. It worths noting that the proposed scheme does batching and reordering in a different fashion than a conventional MIX. Frames are transmitted first-in-first-out on a per each destination basis, but on the node level, frames are transmitted in a different order than when they arrive. The scheme is also very efficient in achieving the security goal. With one broadcast, all neighbors receive a masked data frame. To an unintended receiver, it provides a cover for the node's ensuing transmissions. To achieve the same effect in switching network, multiple transmissions on explicit links to neighbors are needed.

In addition to timing attack, the proposed scheme is also resistant to other attacks. As we mentioned, the padding in a frame's pseudo header must be changed when the frame is retransmitted. This prevents content attack. Size attack is prevented by using fixed-size data frames. Per-hop encryption of frames effectively stops flooding attack.

9.4 Performance Evaluation

In this section, we present the simulation experiments we have carried out to evaluate the performance of our protocol using the Network Simulator, *ns-2* [2]. We present results obtained from experiments in a static wireless ad hoc network which consists of 50 nodes. The radio interface of each node simulates the commercial 914MHz Lucent WaveLAN DSSS radio interface with the transmission range of 250m and the nominal data rate of 2 Mbits/sec. The ns-2 simulator uses the Two-way Ground model to simulate radio signal propagation in open space. In our experiments, nodes are randomly distributed in a 1000m x 1000m square area, and there are 20 CBR connections in the network that generate traffic. The source-destination pairs are randomly chosen from all nodes. The source node of each connection continuously generates data packets of 512 bytes. The average packet generation rate is a parameter that can be varied to control the traffic load. For each connection, a shortest path set is computed at simulation start-up time. Then, when each packet is generated, a path in the set is selected for routing the packet. We do not use a dynamic routing algorithm because we wish to isolate the behavior of our protocol. In each experiment, the simulation run time is 600 seconds. Results are averaged over 10 runs with identical parameter values but different seeds for the random number generator.

In Fig. 9.6, we show the end-to-end data packet delivery fractions under different traffic loads. For comparison purpose, we also show the performance of a "pure" broadcast scheme, i.e., without acknowledgment. We can see that even with light traffic load, the pure broadcast cannot ensure delivery of all frames, and when traffic load increases, its delivery fraction drops fast. At the same time, our scheme achieves significantly higher delivery fractions. The figure also illustrates the effects of the minimal polling set size on the performance. When a larger polling set is required, the duration of the polling process has to be longer, which increases the probability that a data frame is corrupted by hidden nodes' transmissions.

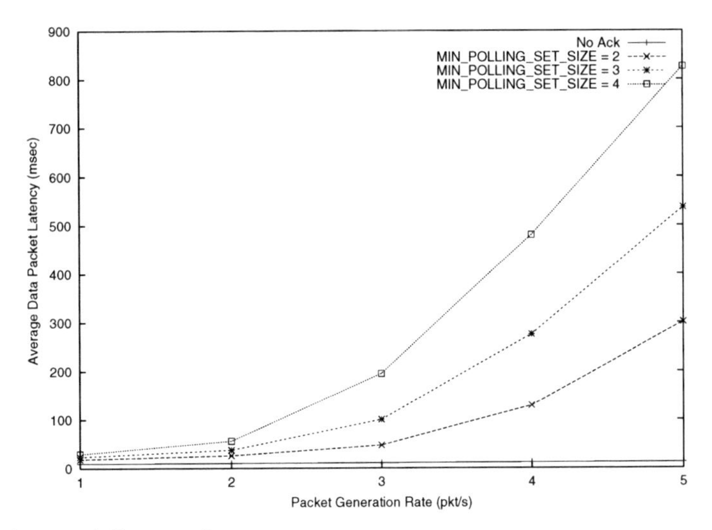

Fig. 9.6. Data delivery ratio

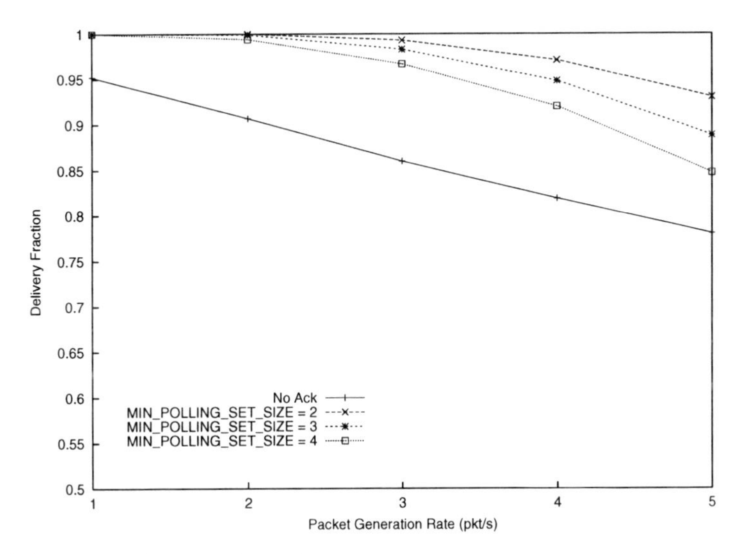

Fig. 9.7. End-to-end data packet latency

In Fig. 9.7, we show the average end-to-end data packet latency under different traffic loads. Since the network is static, there is no routing delay. We also ignore the CPU processing delay at each intermediate node. Therefore, the end-to-end packet latency here includes queueing delays, retransmission delays and propagation delays. It is shown that, on the average, our scheme has much higher packet latency than unreliable, pure broadcast scheme. This is caused by retransmission and batching. When the minimal polling set size increases, the packet latency increases very fast, especially when traffic load is high. The reason is that a larger polling set means higher probability of transmission failure, which makes each node wait for a longer time before next retry. If user's application has delay constraint, a trade-off on security may be needed.

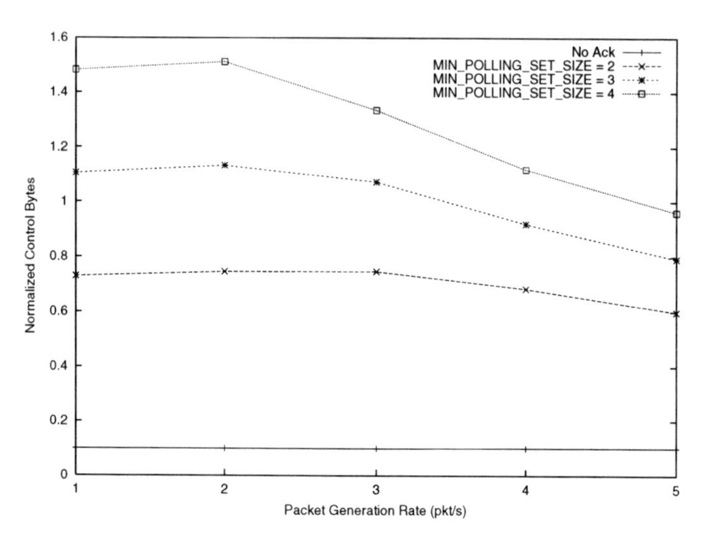

Fig. 9.8. Normalized control bytes

In Fig. 9.8, we show the overhead of our scheme under different traffic loads. We use the metric *normalized control bytes overhead,* which is defined as the total bytes of transmitted control data (POLL, REPLY, MAC header) divided by the total bytes of received data payloads by all nodes. For pure broadcast, this overhead is a constant, equal to the size of a MAC header divided by the size of a MAC frame body. It is shown that the normalized control overhead decreases as the traffic load increases. The reason is that, in this case, there tend to be multiple frames in a node's queue, and each polling process can be followed by multiple data transmissions. In other words, each polling is more efficient. Another observation is that the normalized control overhead is high when the minimal polling set size is large. This is because more dummy POLLs may need to be generated to meet the minimal polling set size constraint.

9.5 Conclusions

In this chapter, we present the design of an anonymous MAC protocol for wireless ad hoc networks. We set two goals for the protocol. One is receiver anonymity. Another is reliability. The former is achieved with link encryption and broadcasting of data frames. The latter is achieved by a selective repeat retransmission scheme, combined with a polling mechanism. We present a security analysis of the protocol and discussed its behavior under different attacks. We also evaluated the performance of the protocol. Simulation results indicate that the protocol increases the packet delivery ratio at a cost of larger packet latency. It is also shown that different trade-offs between the two goals can be achieved by varying a parameter value. This protocol could be incorporated with existing ad hoc routing algorithms (such as DSR [6], AODV [9], DSDV [8]) to provide a good solution for connection anonymity in wireless ad hoc networks.

Acknowledgements

The author of this chapter is extremely grateful to Professor Nitin H. Vaidya of University of Illinois at Urbana-Champaign for inspiring discussions and critical comments during the preparation of this chapter.

References

1. Bahl P, Padmanabhan VN (2000) RADAR: An in-building RF-based user location and tracking system. In: IEEE INFOCOM, pp 775–784
2. Berkeley U, LBL, USC/ISI, Xerox-PARC (2003) NS notes and documentation. http://www-mash.cs.berkeley.edu/ns
3. Boukerche A, El-Khatib K, Xu L, Korba L (2004) A novel solution for achieving anonymity in wireless ad hoc networks. In: ACM Workshop on Performance Evaluation of Wireless Ad Hoc, Sensor and Ubiquitous Networks (PE-WASUN), Venice, Italy
4. Chaum D (1981) Untraceable electronic mail, return addresses, and digital pseudonyms. In: Communications of the ACM, vol 24, no 2, pp 84–88
5. IEEE (1999) IEEE std 802.11, 1999 edition, Wireless LAN medium access control (MAC) and physical layer (PHY) specifications. http://standards.ieee.org/getieee802/802.11.html
6. Johnson D, Maltz DA (1996) Dynamic source routing in ad hoc wireless networks. In: Imielinski I, Korth H (eds) Mobile Computing, vol 353, Kluwere Acedemic Publishers, pp 153–181

7. Kong J, Hong X (2003) ANODR: Anonymous on demand routing with untraceable routes for mobile ad-hoc networks. In: MobiHoc'03, Annapolis, MD, USA
8. Perkins C, Bhagwar P (1994) Highly dynamic destination-sequenced distance-vector routing (DSDV) for mobile computers. In: ACM SIGCOMM'94 Conference on Communications Architectures, Protocols and Applications, pp 234–244
9. Perkins CE (1997) Ad-hoc on-demand distance vector routing. In: IEEE MILCOM
10. Pfitzmann A, Köhntopp M (2000) Anonymity, unobservability, and pseudonymity: a proposal for terminology. IETF Draft, version 0.14
11. Pfitzmann A, Waidner (1985) Networks without user observability - design options. In: EUROCRYPT, vol 219 of Lecture Notes in Computer Science, Springer-Verlag
12. Reed MG, Syverson PF, Goldschlag DM (1997) Anonymous connections and onion routing. In: IEEE Symposium on Security and Privacy
13. Smailagic A, Kogan D (2002) Location sensing and privacy in a context-aware computing environment. In: IEEE Wireless Communications, vol 9, no 10
14. Wu X, Bhargava B (2005) AO2P: Ad hoc on-demand position-based private routing protocol. In: IEEE Transaction on Mobile Computing

10 Hardware/Software Solution to Improve Security in Mobile Ad-hoc Networks

Sirisha Medidi, José G. Delgado-Frias, and Hongxun Liu

School of Electrical Engineering and Computer Science, Washington State University, Pullman, WA 99164-2752

10.1 Introduction

Ad hoc networks are the preferred means of communication where infrastructure is not available in hostile environments for information gathering and time critical decision-making activities. Additionally it would helpful if networks are able to support secure communication while maintaining a high level of network performance. Ad hoc networking opens up a host of security issues, including: (1) *Wireless links are especially vulnerable to eavesdrop.* This may give an adversary access to secret/private information. (2) *Establishing trust among the communicating parties is difficult.* There is no centralized infrastructure to manage and/or to certify trust relationships. This is compounded by the fact these networks are often very dynamic –with nodes free to join and leave at will– and thus having network topology and traffic changing dynamically. (3) *Malicious nodes are difficult to identify by behavior alone.* Many perfectly legitimate behaviors in wireless networking may seem like an attack. (4) *Selfish behavior or node misbehavior is also likely.* Due to node limitations/constraints nodes may opt to go into selfish mode.

Achieving security for ad-hoc networks – To achieve a secure ad-hoc network will undoubtedly require a more comprehensive approach with more sophisticated resources that are integrated into the information-gathering strategies of wireless ad-hoc routing protocols. The proposed approach takes a thorough look at secure wireless ad-hoc networking from a real-time perspective. We propose to incorporate *design for security* (or *design for intrusion-intolerance*) as an integral part of the ad-hoc networks operational specification. The integration includes augmentation of protocols with security and Quality-of-Service (QoS) primitives. Rather than relying on technologies designed for wired networks and currently implemented at the network layers on wireless systems, we believe that multiple strategies are needed to make ad-hoc systems wireless-aware, efficient, and secure.

Handling malicious or unreliable node – There are three steps in handling a malicious node: *detect* malicious behavior, *identify* the malicious node, and *remove* the undesirable node from the network or otherwise cope with it. Ideally techniques to mitigate the effects of malicious or unreliable nodes should: (i) require no modification to protocols, (ii) work with existing routing protocols, (iii) have minimal or no security associations that require the cooperation of other nodes in the network, and (vi) not contribute itself for further attacks on the communication and the routing protocols.

Hardware Monitor – Behavior monitoring by software alone definitely is effective in the detection mechanism. However, false positives could be higher due to the evolving nature of the ad-hoc networks. Furthermore, software based detection can not prevent some malicious nodes from making false indictment to other nodes. Actually, in the mobile ad-hoc environment, it is very difficult to build a trust relationship among the mobile nodes. To have a control on this issue and to further enhance the security of the network, a hardware monitor that provides information to the software layers that is independent of the node's software would be extremely valuable. The hardware monitor can be made tamper resistant and provide trustworthy information to the network. The results provided by the hardware monitor can be used by a reputation system. When there are conflicts between the detection results from the hardware monitor and software monitor, the results sent by the hardware monitor are selected. The hardware monitor should ideally provide the software layers information about: (i) malicious packet drop, (ii) malicious misroute, and (iii) bogus routing information.

Routing problem – Spurious route requests by malicious nodes could cripple the network by introducing broadcast-storm and route-reply storm problems. It is desirable to find a route that has a higher likelihood of surviving over a period of time in spite of node mobility and that has better network resources. Providing routes that are stable based on route statistics could reduce communication disruption time. For effective performance, one needs these features in the routing protocol (all must be energy-efficient): (i) mechanisms to distinguish between false and valid route requests, (ii) ability to adapt to dynamically changing QoS requirements such as battery life, signal strength, bandwidth and latency, and (iii) adaptive mechanisms to detect intrusions and non-cooperative or selfish behavior.

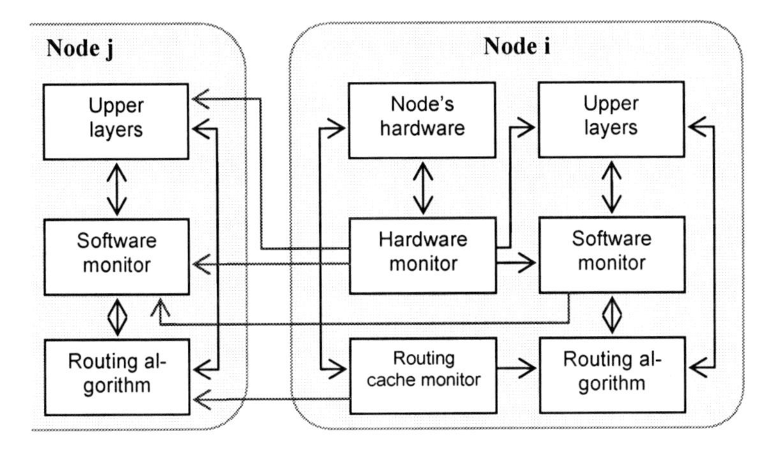

Fig. 10.1. Relationship between software/hardware monitoring and routing

Our Approach – Once undesirable behavior is detected, the malicious nodes will be identified and isolated: doing this leads to secure and QoS-aware routing protocols that strengthen the process of identifying and isolating undesirable nodes. The strength of our approach lies in our ability to incorporate a hardware-monitoring scheme, which is independent of software monitoring techniques. This in turn provides a considerable advantage over existing hardware only or software only techniques. The proposed research aims at developing solutions for misbehavior detection for datagram traffic in addition to the common techniques that are based on TCP (transport control protocol) traffic without any additional security associations that is more common in other solutions. The status information from the hardware monitor will be effectively used in routing decisions to improving the network security as well as the performance.

10.2 Background and Related work

10.2.1 Detection, identification, and isolation of malicious nodes

"Watchdog" [4] is a technique in which each node "snoops" the retransmission of every packet it forwards. If the watchdog detects that a node has not correctly retransmitted the packet, it raises a warning. This requires omni-directional antennas. We developed an unobtrusive monitoring technique, [1, 2, 3] which relies on readily available information at different network levels to detect malicious nodes. The strength of the method is that a single source node can use it without relying on others, making it easy to implement and deploy. Further, there is also no need for security associations between the nodes. Local data such as route request and route error messages, ICMP time exceeded and destination unreachable messages, and TCP timeouts are used to detect misbehavior. Finally, the information is processed to determine if any malicious activity is taking place. In case of undesirable activity, the node is alerted so that it can act. Currently the technique can identify Byzantine faults such as packet drop attack and misrouting. Experiments were conducted using an ns-2 network simulator (details in [1, 2, 3]). The detection effectiveness improves with increase in the percentage of malicious nodes.

We have proposed techniques to improve the performance of nodes in a network by means of novel hardware. This includes buffer schemes that use more efficiently the buffer space in a multiple port node [6]. We proposed an original high-performance cache technique for routing [7, 8, 9]. This technique takes advantage of temporal and geographical locality of packets. T. Chiueh and P. Pradhan [10] proposed to use a conventional cache; this approach has problems with collations due to its associability limitations.

10.2.2 Secure and QoS-aware routing

To achieve optimal availability, routing protocols should be robust against both dynamically changing topology and malicious attacks. Routing protocols proposed so far do not handle security and quality of service within the same protocol. Routing protocols proposed for ad-hoc networks cope well with a dynamically changing topology [11], but none can defend against malicious attacks. We proposed a source-initiated ad-hoc routing protocol (QuaSAR) [12] that adds quality control to all the phases of an on-demand routing protocol. QuaSAR gathers information about battery power, signal strength, bandwidth and latency during route discovery and uses it in route choosing. Also, our approach has proactive route maintenance features in addition to the reactive maintenance. Simulation experiments confirm that QuaSAR performs better than Dynamic Source Routing (DSR) in terms of throughput and delivery ratio [12].

10.3 Comprehensive Software/Hardware Schemes for Security in Ad-hoc Networks

In this section we present our proposed approach to security and QoS in Ad-hoc networks. We have divided this proposed research ideas into two broad categories: (i) Misbehavior detection, identification and isolation of malicious nodes, and (ii) Secure, QoS-aware routing.

10.3.1 Detecting misbehavior, identifying and isolating malicious nodes

10.3.1.1 Software Monitoring

The algorithms we have developed for misbehavior include detection of packet dropping and packet misrouting done offline by analyzing the simulation traces. Algorithms to detect attacks on routing protocols also need to be developed. Techniques such as varying both detection interval and alert threshold will decrease false positives. To further generate triggers for potential attack scenarios or intrusions on the routing protocol, one can use a model-based pattern analysis technique that is loosely based on an expected model of behavior of the routing protocol being used. This can be done modeling the protocol activities as a finite state machine, identifying the sequence of unusual state changes, and getting information from the hardware monitor. Certain learning mechanisms will be incorporated to help with identifications. These techniques will help detect both non-cooperative and selfish behaviors such as nodes that refuse to provide routing service to others (perhaps to conserve battery power) but also ask for and accept service when in need. Experimental results from ns-2 simulations can be used to fine-tune the system. One good way to identify malicious nodes is for each node to initiate the identification process by itself. We can use TCP time out, ICMP destination unreachable message, and route error messages to narrow the malicious node to a set of two nodes. Once the malicious nodes are identified, the source nodes can use this information in their routing decisions.

10.3.1.2 Hardware Monitoring

We propose a novel hardware based node monitoring approach. In this approach a number of monitoring schemes are implemented in hardware. These monitors are kept independent from the software layer of the same node. Even though the software could be compromised by a virus or user, the hardware is made tamper resistant. After the hardware detects the software's misbehavior, it will report such information to other nodes. Upon receiving the warning message, the other nodes can lower the rating the misbehaving node or even isolate the node from the network. The hardware schemes observe traffic through the node, status of queues, and status of neighboring nodes.

The hardware monitor provides information about the node's potential underperformances to its neighbors. The information that is passed on to other nodes includes: packet drop rate above a preset threshold, input queue full rate, and routing modification. Software solutions are usually good at identifying communication paths that may include a malicious node in ad-hoc networks. But, these solutions have problems in pinpointing the exact node that is misbehaving. In these cases, the proposed hardware monitoring technique can help software to identify these nodes and, above all, the potential cause of the problem.

Hardware detects the malicious behaviors through the mechanism called internal monitoring. A hardware monitor observes the behavior of the node's software and reports to neighboring nodes accordingly. When the software layer drops packets, the hardware monitor determines the drop rate and reports this node to other nodes in the same ad-hoc network if the drop rate reaches a pre-defined threshold value. The assumption is that all the mobile nodes have the proposed hardware.

The implementation of internal monitoring is through an adaptive drop counter that records the packet-dropping rate of the software layers. The counter records the number of packets dropped during a given period of time. If the counter reaches or exceeds a threshold value, a reporting mechanism is triggered. Both the period of time and the threshold are adaptive. They can be adjusted according to the traffic and other factors. For example, the detection period could be shortened for a heavy burdened node. A concrete implementation is described in the following two-timer scheme which is based on DSR.

In the two-timer scheme, there are two timers, Detection Timer and Reward Timer, in addition to the drop counter. The Detection Timer (DeT) is used to detect if the wireless node forwards a received packet during the detection period. A received packet will first arrive at the hardware layer and trigger this timer. Then the packet will be passed to the software layer of the same node. The software layer of a good-behaving node will process the packet and forward the packet according to the content of the packet. If the node never forwards the packet and DeT expires, the value of drop counter is increased by 1. If the node's DeT keeps incrementing the counter, the drop counter of a misbehaving node will reach a predetermined threshold, thus triggering the warning message to be sent.

The other timer, Reward Timer (RwT), is used to reward the good-behaving node during the route discovery process of DSR. During the process of route discovery, the route request packet is usually sent through broadcast, which will cause the same route request packet to be received several times by a single node. As shown in Fig. 2, when node A receives a route request packet and broadcasts that packet, its neighbor B will receive and broadcast the packet. Due to the nature of broadcast, Node A will receive the same route request packet again from node B. If node A has a few neighbors within its transmission range, it is likely that A will receive a few duplicate route requests. To compensate for such irregularity, after a good-behaving node forwards a route request, the node will be rewarded a grace period by means of Reward Timer (RwT) during which the node could "drop" the duplicate route request without being penalized. As shown in Fig. 3, a wireless node receives a route request packet A (RRP_A) which starts the DeT. During the period of DeT, the RRP_A is forwarded, which starts the RwT. Then the three duplicate RRP_A received during RwT will not be counted as new packets and dropping the three duplicate RRP_A will not increase the drop counter. On the other hand, if the RRP_A is not forwarded during the period of DeT, there will not be RwT. Then the duplicate RRP_A will be counted as new packets and the dropping of these duplicate RRP_A will increase the drop counter.

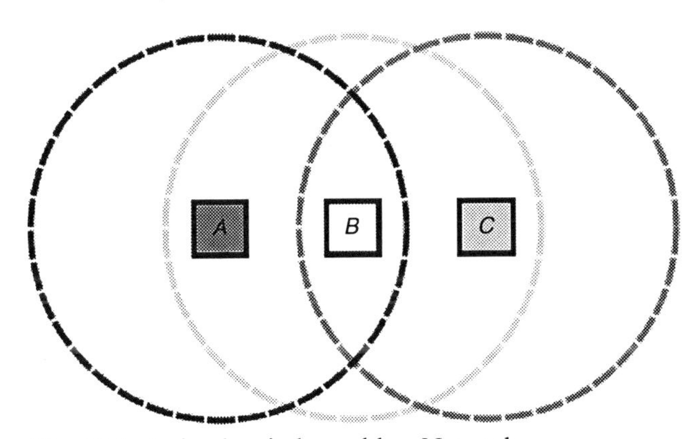

Fig. 10.2. A simple wireless ad-hoc Network

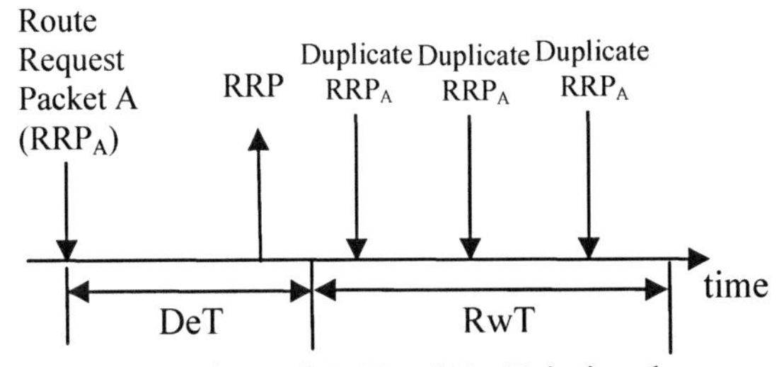

Fig. 10.3. The functions of DeT and RwT during the processing of route request packet

A number of simulations are run to evaluate the performance of two-timer scheme. Two metrics are used in the evaluation process: *Detection Effectiveness* and *False Positive*. *Detection Effectiveness* measures how well the two-timer scheme performs in identifying the misbehaving nodes. The detection effectiveness is measured as follows.

$$DetectionEffectiveness = \frac{Detected\_nodes}{Total\_misbehaving\_nodes} \times 100$$

In the formula above, detected_nodes are the misbehaving nodes that have been identified. For example, if the scheme detects all of the misbehaving nodes in the network, the detection effectiveness is 100%. On the other hand, if it cannot detect any misbehaving node, the detection effectiveness is 0%.

Some good-behaving nodes may be misidentified as misbehaving nodes; this is called *False Positive*. It is measured as the number of good but detected misbehaving nodes divided by the total number of detected misbehaving nodes.

$$FalsePositive = \frac{missclassified\_missbehaving\_nodes}{Detected\_nodes} \times 100$$

The detection object would be to have 0% false positive and 100% detection effectiveness. But this in turn is extremely difficult to obtain. Our goal in this study is to keep false positive at 0% and try to get the detection effectiveness as high as possible. Having a number of *good* nodes being misclassified has a negative effect on the overall performance of the network. On the other hand, some nodes that have the potential of misbehaving may not be involved in any communication path. This in turn makes those nodes difficult to detect and having minimum impact on the network performance. Thus, our goal is to get very low false positive, followed by high detection effectiveness.

In the simulations, the percentage of misbehaving nodes ranges from 10% to 40%, with an increment of 10%. For each percentage of misbehaving nodes, 100 times of simulations are run. The simulation results are shown in both Table 1 and Fig. 4.

Table10.1. Simulation results

Detected Misbehaving Nodes (%)	Percentage of Misbehaving Nodes			
	10%	20%	30%	40%
0-60	8	6	4	6
60-70	5	0	5	3
70-80	0	2	5	3
80-90	12	7	14	12
90-100	0	15	16	21
100	75	70	56	55

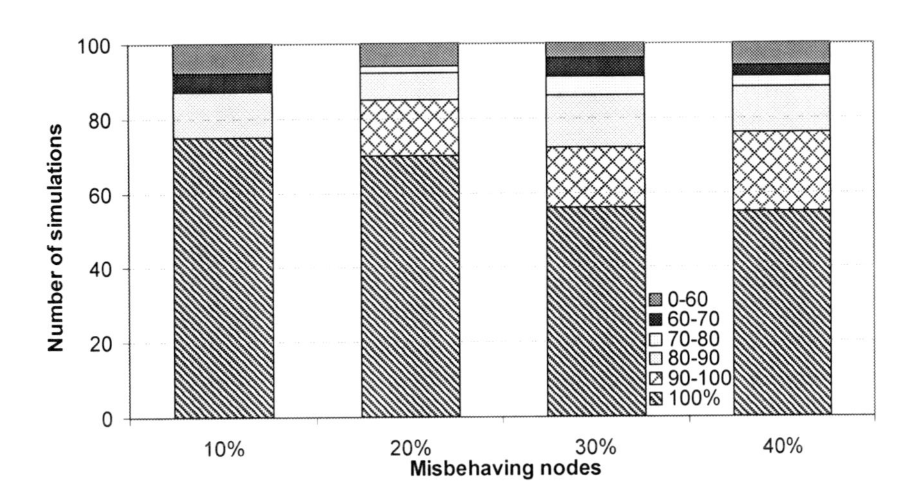

Fig.10.4. Simulation results with 0% false positive

In Fig. 4, there are four columns. Each column corresponds to a percentage of misbehaving nodes in the simulation. As mentioned earlier, for each percentage of misbehaving nodes 100 simulations are run. As shown in Fig. 4, the scheme can detect 80% or above misbehaving nodes in all the percentages at a probability of almost 90%. For all the percentages, they have perfect detections at a probability of at least 55%. An interesting phenomenon is that the probability of perfect detection decreases as the percentage of misbehaving nodes increases. One reason is that more misbehaving nodes lead to more separated networks. In such cases, there is no enough traffic to pass the misbehaving nodes while the two-timer scheme needs a relatively connected network. Another reason is that as the percentage of misbehaving nodes increases the number of packets resent increases, leading to longer response time for the good-behaving nodes.

The two-timer scheme proposed has the following features:

High detection of misbehaving nodes – The proposed scheme can detect 80% or above misbehaving nodes with a probability of almost 90%.

Zero false positives – The two timers are set to achieve a 0% misclassification of good-behaving nodes as misbehaving nodes.

Minor changes to software layer – The proposed scheme requires very little change to the present software layer and can be easily implemented at the hardware layer due to the simple nature of the scheme.

The simulation results show that the two-timer scheme can detect misbehaving nodes accurately in terms of detection effectiveness and false positive. Currently, we are conducting some enhancements to the two-timer scheme and designing other more efficient hardware scheme to detect packet dropping.

Another hardware monitor checks the input buffer to determine the time that this buffer is full. This is an important issue since packet dropping may be due to lack of memory resources. If the time that the buffer is full is higher than a threshold value, the hardware will report this to other nodes. This in turn will indicate to other nodes that the current node can not keep up with handling too many packets and it is not a malicious node.

Besides the packet dropping monitor and buffer monitor, the misrouting monitor is responsible to detect packet misrouting. For example, a few misbehaving nodes can collude with each other and send forwarding packets to each other by modifying the routing information inside the packets. Detecting colluding nodes using software scheme usually time consuming and very complicated. Using hardware monitor to detect misrouting can relieve the software from such complicated work. The misrouting monitor needs to detect such misbehavior and report the detection results to other nodes. The other nodes can use the information received to isolate the colluding nodes from the network. Currently, we are working on combining misrouting monitor with the packet dropping monitor.

10.3.1.3 Software/Hardware Monitoring

The software monitoring will enable us in detecting, identifying and isolating malicious nodes. Through the help of hardware monitoring, the software layer will be assisted to make a more precise determination of malicious nodes and the causes of potential problems. The software layer will determine the actions that need be taken to avoid malicious nodes and to improve throughput, quality of service, and/or reliability.

It should be pointed out that hardware flow monitor makes no decisions rather it provides independent information to its own node and adjacent nodes. Novel algorithms are going to be developed that take into account this additional information. Since there is a new independent source of information, the new algorithms for detecting, identifying and isolating malicious nodes will be more precise with far fewer false positive outcomes. Our groundwork on this project has yielded extremely positive results that need be fully studied and integrated in the proposed research.

10.3.2 Secure, QoS-aware routing

10.3.2.1 Software Techniques

To achieve IETF (Internet Engineering Task Force)-compatible protocol specification of the secure routing, we propose extensions of DSR that encapsulate source routing capabilities, but with minimal changes and overhead. Messages such as route request (RREQ) and route reply (RREP) need to be augmented to reflect the malicious nodes or suspicious activity by the nodes in the path, and also quality of service requirements. Above and beyond format specification, a key technical challenge lies in managing RREQ implosion (the "broadcast storm" problem). Some of the techniques we employed in quality of service routing [12] can apply to secure routing. A second issue is the route reply storm problem that is created due to the number of routes that are sent back to the source. Selective route replies that we developed in [12] can be adapted to alleviate this problem. A third issue is that there needs to be a proactive mechanism to preempt route breaks arising due to signal strength weakening (when the mobile node moves out of range), battery power depletion, and memory shortage (node becomes selfish and drops packets). One way to address this is to send a route change request (RCR) to find a new route. In [13], a proactive mechanism is proposed to preempt route breaks based on signal strength measurements. This idea can be enhanced to also include route breaks due to low battery power and memory shortage. Finally, one can incorporate learning mechanisms in the routing process to detect intrusions including spurious route requests and non-cooperative or selfish behaviors. The knowledge gained through our misbehavior detection and identification process will be integrated with the routing decisions to further improve the routing performance. Testing and refining these protocols and algorithms in an actual ad-hoc network test-bed would provide us insight into how the proposal works.

10.3.2.2 Hardware Support

Routing cache monitor is another innovative technique to observe and report changes in the routing. As a routing path is established, information about this path is inserted in a cache memory. As packets for this path pass through the node, the cache checks packet forwarding. If the routing is changed, this may trigger a reporting mechanism of a potential problem. Our cache technique takes advantage of temporal and geographical locality of the packets [8]. When bogus routing information is reported, the routing protocol incorporates this into its routing decisions. We anticipate that using this additional information will further enhance the security and performance of the network.

10.4 Implications and Future Research

In this paper, we make a case that to realize secure communication in ad-hoc networks, it is necessary to develop comprehensive techniques to detect, identify and isolate malicious nodes in the network and then integrate this information into routing decisions. Based on our preliminary results and our experience, we believe such integration would not only improve the security of the network but also its performance. In our experience, software only solutions have given us good detection effectiveness in terms of malicious behavior detection and reasonable false positive level. Providing an independent source of monitoring with hardware integrated into the software layers would greatly reduce the false positives and increase the detection effectiveness of our techniques. We have included a simple, but yet powerful, two-timer scheme. This scheme is capable of detecting a large percentage of misbehaving nodes while keeping false positive detection to zero. Further using route-cache monitor would greatly enhance routing security. This multi-layer hardware/software approach will significantly enhance the security and performance of mobile ad-hoc networks.

As explained in this paper, we have came to the conclusion that having two independent monitors (software and hardware monitors) could lead to a significant enhancement of security and performance of mobile ad-hoc networks.

References

1. S. Medidi, M. Medidi, S. Gavini (2003) Detecting Packet Dropping Faults in Mobile Ad-hoc Networks. In *Proc. of IEEE ASILOMAR Conference on Signals, Systems and Computers*, volume 2, pp. 1708–1712
2. S. Medidi, M. Medidi, S. Gavini, R. L. Griswold (2004) Detecting Packet Mishandling in MANETs. In *Proc. of Security and Management Conference*, pp. 40–44
3. R. L. Griswold, S. Medidi (2003) Malicious Node Detection in Ad-hoc Wireless Networks. In *Proc. SPIE AeroSense Conference on Digital Wireless Communications*, volume 5100, pp. 40–49
4. S. Marti, T. J. Guili, K. Lai, M. Baker (2001) Mitigating routing misbehavior in mobile ad hoc networks. In *Proc. of ACM SIGCOMM*, pp. 255–265
5. S. Buchegger, J. Y. Le Boudec (2002) Nodes bearing grudges: Towards routing security, fairness, and robustness in mobile ad hoc networks. In *Proc. of the Parallel, Distributed and Network-based Processing*, pp. 403–410
6. J. Liu, J. Delgado-Frias (2005) DMAQ Self-Compacting Buffer Schemes for Systems with Network-on-Chip. In *Proc. of Int. Conf. on Computer Design*, pp. 97–103
7. J. Nyathi, J. G. Delgado-Frias (2002) A Hybrid Wave-Pipelined Network Router. In *IEEE Transactions on Circuits and Systems*, 49(12): 1764–1772
8. J. J. Rooney, J. G. Delgado-Frias, D. H. Summerville (2004) An Associative ternary cache for IP routing. In *Proceedings of IEE Section E: Computers and Digital Techniques*, volume 151, pp. 409–416
9. D. H. Summerville, J. G. Delgado-Frias, S. Vassiliadis (1996) A Flexible Bit-Associative Router for Interconnection Networks. In *IEEE Transactions on Parallel and Distributed Systems*, 7(5): 477– 485
10. T. Chiueh, P. Pradhan (2000) Cache Memory Design for Internet Processors. In *6th Symposium on High Performance Computer Architecture (HPCA-6)*, Toulouse, France
11. D. B. Johnson, D. A. Maltz, Y. C. Hu, J. G. Jetcheva (2003) The dynamic source routing protocol for mobile ad hoc networks (DSR). Internet draft, www.ietf.org/proceedings/03mar/I-D/draft-ietf-manet-dsr-08.txt
12. S. Medidi, K. Vik (2004) QoS-Aware Source-Initiated Ad-hoc Routing. In *Proc. of IEEE Conference on Sensor and Ad Hoc Communications and Networks*, pp. 108–117
13. T. Goff, N. Abu-Ghazaleh, D. Phatak, R. Kahvecioglu (2001) Preemptive routing in ad hoc networks. In *Journal of Parallel and Distributed Computing*, 63(2): 123–140

Index

Printed in the United States
99049LV00003B/226/A